Springer Series in Reliability Engineering

Series Editor

Professor Hoang Pham
Department of Industrial and Systems Engineering
Rutgers, The State University of New Jersey
96 Frelinghuysen Road
Piscataway, NJ 08854-8018
USA

Other titles in this series

System Software Reliability
Hoang Pham

Reliability and Optimal Maintenance
Hongzhou Wang and Hoang Pham

Applied Reliability and Quality
B.S. Dhillon

Shock and Damage Models in Reliability Theory
Toshio Nakagawa

Risk Management
Terje Aven and Jan Erik Vinnem

Satisfying Safety Goals by Probabilistic Risk Assessment
Hiromitsu Kumamoto

Offshore Risk Assessment (2nd Edition)
Jan Erik Vinnem

The Maintenance Management Framework
Adolfo Crespo Márquez

Human Reliability and Error in Transportation Systems
B.S. Dhillon

Complex System Maintenance Handbook
D.N.P. Murthy and Khairy A.H. Kobbacy

Recent Advances in Reliability and Quality in Design
Hoang Pham

Product Reliability
D.N.P. Murthy, Marvin Rausand and Trond Østerås

Mining Equipment Reliability, Maintainability, and Safety
B.S. Dhillon

Advanced Reliability Models and Maintenance Policies
Toshio Nakagawa

Justifying the Dependability of Computer-based Systems
Pierre-Jacques Courtois

Reliability and Risk Issues in Large Scale Safety-critical Digital Control Systems
Poong Hyun Seong

Failure Rate Modeling for Reliability and Risk
Maxim Finkelstein

The Complexity of Proceduralized Tasks
Jinkyun Park

Risks in Technological Systems
Göran Grimvall, Åke J. Holmgren, Per Jacobsson and Torbjörn Thedéen

Maintenance for Industrial Systems
Riccardo Manzini, Alberto Regattieri, Hoang Pham and Emilio Ferrari

Mine Safety
B.S. Dhillon

The Complexity of Proceduralized Tasks
Jinkyun Park

Simulation Methods for Reliability and Availability of Complex Systems
Javier Faulin, Angel A. Juan, Sebastián Martorell and José-Emmanuel Ramírez-Márquez

Reliability and Safety Engineering
Ajit Kumar Verma, Srividya Ajit and Durga Rao Karanki

Albert Myers

Complex System Reliability

Multichannel Systems with Imperfect Fault Coverage

2nd Edition

Springer

Albert Myers
Myers Consulting
835 Reposado Drive
La Habra Heights
CA 90631
USA
al@amyersconsulting.com

ISSN 1614-7839
ISBN 978-1-4471-5713-7 ISBN 978-1-84996-414-2 (eBook)
DOI 10.1007/978-1-84996-414-2

British Library Cataloguing in Publication Data
A catalogue record for this book is available from the British Library

First edition published by Myers Consulting and M4 LLC, ISBN 978-0-6152-1592-1

Cover design: eStudioCalamar, Girona/Berlin

Printed on acid-free paper

Springer is part of Springer Science+Business Media (www.springer.com)

Preface

This book focuses on reliability modeling of complex multichannel systems, a well-known example of which are digital fly-by-wire aircraft control systems. Since the consequences of failure of these systems are severe, having both substantial economic and personnel safety implications, it is of critical importance that the analysis of these systems be done correctly. With the current widespread use of this type of system (even automotive "drive-by-wire" systems are now being seriously considered), correct assessment of the reliability of such systems has become increasingly important. Not only is a correct reliability model crucial for understanding the system once it is fully designed, but it also serves as a critically important tool in the assessment of the myriad design alternatives that are considered during the design phase.

Despite the importance of correctly modeling these complex multichannel systems, there is a paucity of literature addressing the topic; this is especially true of the reliability assessment of redundant systems that use voting-based selectors that may be subject to imperfect fault coverage. All redundant systems must have some means of selection among their redundant inputs, a task that has been termed *redundancy management* (in the aerospace vernacular, at least). Redundancy management can seldom, if ever, be done with perfect certainty, and therefore, redundant systems are subject to imperfect fault coverage. Imperfect fault coverage has a significant adverse impact on the reliability of redundant systems (as compared with systems that have perfect fault coverage) and, as a result, cannot properly be ignored in the assessment of complex multichannel system reliability.

Even basic complex system reliability modeling (with perfect fault coverage) is intrinsically difficult, and it is a well-known example of an NP-complete problem. The correct modeling of redundant systems, when accounting for the effects of imperfect fault coverage, requires the use of powerful analysis tools. Historically, the analysis of multichannel system reliability that accounts for voting-based imperfect fault coverage required the development of very complex conditional probability models. These models were difficult and tedious to construct and, because of this difficulty, required a great deal of additional effort to validate. Consequently, they tended to play a limited role in the initial design phases of a system and were primarily used only to assess the reliability of the final product.

Analysis techniques and tools now exist to correctly assess complex multichannel systems both quickly and accurately; these techniques and tools are fully explained and their use demonstrated in this book. The techniques discussed here include the use of binary decision diagrams (BDD) and BDD-based algorithms for the reliability assessment of redundant systems subject to imperfect fault coverage. The objective of this book is to provide a set of basic analytical and numerical techniques that are suitable for modeling these systems. The approach of the book is to concentrate on the demonstration of these techniques, rather than on the development and derivation of their underlying theoretical basis.

This book provides the necessary background for an engineer to develop valid reliability models for large, complex redundant systems, including those subject to imperfect fault coverage.

Los Angeles, California
May 2009

Albert Myers

Contents

Notation

n	Number of redundant elements in a k-out-of-n:G system
k	Number of operational elements in a k-out-of-n:G system
p	Reliability of i.i.d. redundant elements
$\mathbf{p}$	Set of redundant system element reliabilities
$\mathbf{q}$	Set of redundant system element unreliabilities
i	Label for redundant element i
p_i	Reliability of redundant element i (not necessarily i.i.d.)
q_i	$(1 - p_i)$; unreliability of redundant element i (not necessarily i.i.d.)
c	Coverage of the one-on-one fault for OLC model
c_i	Coverage of the ith component for ELC model or coverage of the ith failure for FLC model
t_M	Mission time
$\mathbf{c}$	Set of covered failure probabilities for FLC or ELC reliability models
$\lvert\mathbf{x}\rvert$	Number of elements in set $\mathbf{x}$ for any set $\mathbf{x}$
P(•)	Probability of •
R	Function defining the reliability of an at least k-out-of-n:G system
$\mathbf{pT}(i, \mathbf{p})$	Set of products of the k-subsets of $\mathbf{p}$ with exactly i elements
$\mathbf{qT}(i, \mathbf{q})$	Set of products of the k-subsets of $\mathbf{q}$ with exactly $(n - i)$ elements
$\mathbf{cT}(i, \mathbf{c})$	Set of products of the k-subsets of $\mathbf{c}$ with exactly $(n - i)$ elements ($\mathbf{c}$ = ELC coverage vector)

Abbreviations

BIT	Built-in test
BDD	Binary decision diagram
BSV	Bernoulli state variable
CCDL	Cross-channel data link
CPM	Conditional probability model
DFBW	Digital fly-by-wire
ELC	Element level coverage
ERL	Effective redundancy level
FCASE	Flight Critical Aircraft System Evaluation
FDT	Fault detection threshold
FLC	Fault level coverage
IFC	Imperfect fault coverage
MVS	Mid-value-select (redundant component voting)
OLC	One-on-one level coverage
PFC	Perfect fault coverage
RM	Redundancy management
SDP	Sum of disjoint products
SOS	Sum over states
dpd	Decades per decade (slope of curve on a log-log plot)
fpmh	Failures per million hours
i.i.d.	Independent and identically distributed

Chapter 1
Introduction

Abstract This chapter provides brief introductory comments on two issues of particular significance in assessing the reliability of redundant multichannel systems: imperfect fault coverage and computational complexity.

1.1 Imperfect Fault Coverage

Critical fault-tolerant systems frequently must use redundancy if they are required to meet extremely high reliability levels. Redundancy management (RM) is the process by which a redundant system selects among its redundant components so as to provide fault tolerance in the event a redundant component should fail. RM consists of the following tasks: detection (recognizing that a fault has occurred among the redundant components); isolation (identifying, given a fault has occurred, which of the components has failed; and reconfiguration (correctly altering the system's behavior so as to prevent a failed component from adversly affecting the system's ongoing performance).

The probability that these RM tasks are accomplished correctly is called *fault coverage* or simply *coverage*. The RM task can seldom be done with perfect certainty; that is, there will be some, (perhaps very small) probability that at least one of the tasks (identification, isolation or reconfiguration) will not be done correctly. Systems that are subject to some level of uncertainty in the RM process are subject to imperfect fault coverage (IFC), whereas systems with a perfect RM process have perfect fault coverage (PFC).

The reliability literature has adequately addressed the assessment of redundant system reliability for the PFC case [1]. The technique for the determination of IFC reliability depends on the nature of the system's RM approach. There are two broad categories of RM design: if a particular component has a given coverage value, the reliability can be modeled using what is termed *element level coverage* (ELC); on the other hand, if the redundant component's coverage is dependent on the fault sequence within the redundant set (first failure, second failure, an so on), the reliability is modeled using *fault level coverage* (FLC). If the redundant components each have an associated self monitoring or built in test capability that is used as the primary basis for RM, these are properly modeled as ELC systems. If, however, the system RM is based on "voting" the redundant components using an approach such as mid-value-select (MVS) when at least three components are potentially available, the system should be modeled as an FLC system. Techniques for modeling ELC type systems have been presented in the literature for some time [2, 3], but FLC systems have only been addressed more recently [4–6].

Imperfect fault coverage, even in circumstances where coverage is quite high, (for example, > 99%), can have a *substantial* impact on system reliability, and it is essential to use the appropriate model. Subsequent chapters discuss how to assess redundant system reliability using a variety of algorithms and computational approaches for PFC, ELC and FLC systems.

1.2 Computational Complexity

Reliability modeling of large, complex systems is an inherently difficult task; the fundamental reliability problem is well known as being NP-complete.[1] Furthermore, correct techniques for modeling redundant systems subject to imperfect fault coverage are not widely known, particularly in the case of systems that use voting as a critical means of redundancy management.

Although the techniques discussed in this book involve a number of basic elements of reliability theory (as well as some more advanced elements), the book is not intended to provide a foundation in the underlying theory. Rather, it is intended to provide engineers with the requisite tools for effectively designing reliable systems. Consequently, only a limited development of the theory is provided, and the computational techniques are often presented without any theoretical justification. Readers seeking a more complete presentation of the foundational elements of reliability theory are encouraged to consult the references provided in the text.

As mentioned, the reliability modeling of large, complex systems is an inherently difficult task, with the fundamental reliability problem being NP-complete.[2] A problem belongs to the NP-hard complexity class when it is more difficult to solve than those problems that can be solved in polynomial time by a nondeterministic Turing machine (and, accordingly, by any processor). A problem is NP-complete when it both belongs to the class NP and is NP-hard. NP-complete problems are the most difficult problems in the class NP [7, 8]. As a practical matter, this means that the time required to solve a general problem of the class NP-hard, at least in

[1] A problem with a computational complexity of $O(n)$ or less is easily solved, but one with a complexity of $O(2^n)$ is very hard to solve, except for small n. Between these two types of problems lie those that can be solved in polynomial time; NP-complete problems belong to this class. If a problem is NP-complete, then there exists no known efficient solution to the general problem (that is, a solution that can be completed in less than polynomial time).

[2] NP stands for nondeterministic polynomial time.

the worst case, increases at a rate that is greater than some polynomial expression given in terms of the problem size. This does not mean, however, that there is no effective approach for the analysis of large systems that are of practical interest to the system designer. This book provides effective, state-of-the-art tools capable of yielding accurate results for complex systems of "real-world" size.

1.3 Symbolic Algebra

Applications, such as *Mathematica*© [9], are capable of performing symbolic algebra operations on large expressions on a desktop computer. These applications can be used to develop exact results for complex systems, including those subject to imperfect fault coverage. Chapters 3 through 5 cover the use these tools to assess the reliability of complex redundant systems.

Mathematica is used extensively in this book to perform both numerical and symbolic evaluations. Although any mathematical analysis package capable of symbolic manipulation could have been used, the present author's opinion is that *Mathematica* is well-suited to perform the kind of analysis outlined in the book. The reader may elect to use a different software package, such as *Maple*, and should not encounter any great difficulty in adapting the techniques, algorithms and functions presented here.

Although all of the fundamental functional and algorithmic definitions are also presented using conventional mathematical notation, the reader is assumed to have a level of *Mathematica* understanding sufficient to follow some of the example calculations. Readers can gain proficiency with the moderate level of *Mathematica* used in this book by following the informative tutorial that is provided with the software package. Additionally, a number of available books introduce the reader to *Mathematica*; among these, the author has found [10–12] to be helpful.

The *Mathematica* examples included in this book are typeset using a feature of *Mathematica* that allows its input and output statements to be saved as a TEX file. The appearance of the results is similar to that of *Mathematica*, with input statements appearing in boldface and the results appearing in the normal font. The typeset fonts are somewhat different, however, from the normal *Mathematica* output. For example,

x = (1 + y)(1 + z)//Expand

$1 + y + z + yz$

Although *Mathematica* is capable of using subscripted variables, their use in general symbolic calculation requires more-advanced programming. For this reason, several of the examples use the notation $\mathbf{p} = \{p1, p2, p3, p4\}$ to represent the elements of a vector **p**, instead of the more conventional notation $\mathbf{p} = \{p_1, p_2, p_3, p_4\}$.

The *Mathematica*-based approach taken in this book is not immune from the "combinatorial explosion" that makes the reliability problem NP-complete. Consequently, the size of the system that can be analyzed in a direct fashion is limited, where the size is measured by the number of independent elements composing the full system. Under most circumstances, this "limit" is in the range of 20 to 25 independent components. In many cases, however, significantly larger system models can be formulated using conditional probability models whose constituent parts have been developed using the more direct techniques. Even within a problem size constraint of 20 to 25 components, many complex systems and subsystems that are difficult to study using other approaches can nevertheless be evaluated as exact symbolic expressions and studied fully within the limits of the *Mathematica*-based approach.

1.4 Binary Decision Diagrams

Regardless of the power of the *Mathematica*-based approach, an understanding of most multichannel systems ultimately requires an analysis of significantly more than 20 to 25 components, and in some cases, constructing a correct conditional probability model may be very difficult. This book also describes techniques that are fully capable of providing exact reliability results for these larger systems as well as for smaller systems. The process of evaluating large systems requires algorithms that are based on binary decision diagrams, which, during the last decade, have been established as the state-of-the-art technique for the assessment of large-system reliability. The binary decision diagram techniques presented in this book have been implemented in the *FCASE* reliability analysis program, which is a code written in ANSI standard C. Documentation for the *FCASE* input syntax is also presented.

References

1. Barlow RE, Proschan F (1975) Statistical Theory of Reliability and Life Testing: Probability Models. Holt, Reinhart and Winston, New York
2. Dugan JB, Trivedi KS (1989) Coverage modeling for dependability analysis of fault-tolerant systems. IEEE Trans Comput 38:775–787
3. Trivedi KS (2002) *Probability and Statistics with Reliability, Queing and Computer Science Applications, 2nd ed.* Wiley, New York
4. Myers AF (2007) *k*-out-of-*n*:G System Reliability With Imperfect Fault Coverage. IEEE Trans Relia 56:464–473
5. Myers A, Rauzy A (2008) Efficient Reliability Assessment of Redundant Systems Subject to Imperfect Fault Coverage Using Binary Decision Diagrams. IEEE Trans Relia 57:336–348
6. Myers AF, Rauzy A (2008) Assessment of redundant systems with imperfect coverage by means of binary decision diagrams. Reliab Eng Syst Saf 93:1025–1035
7. Atallah (ed) (1999)Algorithms and Theory of Computation. CRC Press, USA
8. Dictionary of Algorithms and Data Structures [http://www.itl.nist.gov/div897/sqg/dads/HTML/npcomplete.html]. US NIST. Accessed 27 December 2009
9. Wolfram S (1999) The *Mathematica* Book, 4th edn. Wolfram Media/Cambridge University Press

10. Ruskeepaa H (1999) *Mathematica* Navigator, Graphics and Methods of Applied Mathematics. Academic Press, USA
11. Abell ML and Braselton JP (1997) *Mathematica* by Example. Academic Press, USA
12. Ruskeepaa H (1999) *Mathematica* A Practical Approach, 2nd edn., Prentice Hall, USA

Chapter 2
Basic Elements of System Reliability

It is difficult to get where you want to go if you don't know where that is.

Abstract This chapter presents the basic principles and functional relationships used for reliability assessment of systems with simple interconnections. Systems with simple interconnections are those that can be reduced to a single equivalent element or block through a sequence, however complex, of series and parallel reductions. The analysis of systems with complex interconnections is treated in Chapter 3. Although most multichannel systems have complex interconnections, many fundamental principles of redundant system reliability can be developed and understood using the principles discussed in the present chapter. The objective of this chapter is to first determine the overall reliability of a system that is composed of multiple subsystem elements or components, each with known reliability, and then to determine the reliability of the system that comprises these components.

2.1 The Reliability Function

Reliability is defined as the probability that an element (that is, a component, subsystem or full system) will accomplish its assigned task within a specified time, which is designated as the interval $t = [0, t_M]$. This book deals only with systems consisting of elements that can take on one of two states: either the element is operational (designated as the 1 state) or the element has failed (designated as the 0 state). Furthermore, the book considers only coherent systems, which have the following characteristics: (a) the reliability of the system increases if the reliability of its components increases, and (b) the system has no irrelevant components. The failure of any component or set of components in a coherent system cannot cause an increase in reliability, and every component has some effect, however small, on the overall reliability.

If the reliability of the ith component of a system is p_i and if this component has an unreliability q_i, then

$$p_i = 1 - q_i \tag{2.1}$$

and

$$q_i = 1 - p_i \, . \tag{2.2}$$

Also, since the component is always in one of the two possible states (operational or failed),

$$q_i + p_i = 1 \, . \tag{2.3}$$

To perform quantitative system reliability analysis, it is necessary to ascribe a probability that the individual components p_i either are operational or have failed. A *reliability function*, also called a *survivor function*, defines the probability that the component will perform its intended task (usually subject to some stated set of environmental conditions, such as vibration and temperature) for some specified performance period. The performance period may be a function of cycles, distance or time. Although the techniques presented here can employ a reliability function that depends on any of these three parameters, the focus is on determining the probability of system failure as a function of time. Additionally, although the estimation of system element reliabilities is outside the scope of this book, it is essential that the failure characteristics of the elements be determined using an approach that appropriately accounts for the environment in which they operate. It is also critical that the reliabilities are estimated in a legitimate and appropriately conservative fashion.

Several different functions have been used to characterize the probability distribution of failures as a function of time. Some of the more common reliability functions include the exponential, normal, log-normal and Weibull distributions. In this book, however, the exponential probability distribution is used almost exclusively.[1] The exponential distribution is appropriate for components with a failure rate that is time independent. Most electronic devices demonstrate such a constant failure rate during their useful lifetime, which is the time following a "burn in" that eliminates any weak or faulty components. The reliability function for a single-component system associated with the exponential distribution is

$$r(\lambda, t) = e^{-\lambda t} \, , \tag{2.4}$$

where $r(\lambda, t)$ is the probability that a component with failure rate λ will be operational at time t. The *Mathematica* function implementing Equation (2.4) is

```
r[λ_, t_]:=e^(-λt);
```

This book typically depicts system reliability graphically using log-log plots of the probability of failure as a function of time. Figure 2.1 shows the probability of failure for components with failure rates λ = 100, 200 and 400 failures per million hours (fpmh).

Figure 2.1 illustrates several noteworthy points. Obviously, the probability of component failure increases with time and with λ. Note that the curves, when shown on a log-log plot, are nearly straight lines in the time range of interest. Also, the

[1] The techniques illustrated in this book, however, are not limited to use of the exponential distribution; substitution of another distribution in lieu of the exponential is perfectly valid.

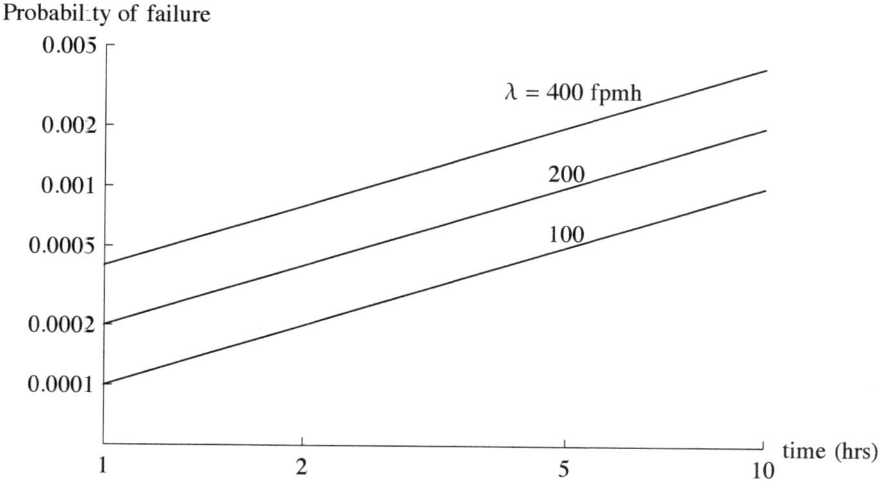

Fig. 2.1 Probability of component failure for various failure rates, λ

probability of failure curves increase uniformly by equal amounts for each doubling of the failure rate, λ. The slope of these curves is approximately one decade of unreliability per decade of time, which is a general feature of all simplex systems with a high degree of reliability over the time span of interest.[2] It is shown later that the slope of a system unreliability curve is related to the level of system redundancy, with the slope increasing as the level of redundancy increases.

2.2 Reliability Functional Block Diagrams

To support the analysis of system reliability, the analyst should first, after careful study of the entire system, depict the overall system design in the form of a reliability functional block diagram.[3] The purpose of the block diagram is to describe the system at its simplest level, while still retaining all of the significant subsystem or component failure information, and to describe the effect that these failures have on the overall reliability of the system. Generally, this means that the functional block diagram represents the system as a collection of "black boxes," each of which is subject to independent failure with respect to the other system elements. Note that

[2] Obviously, since $p = e^{-\lambda t}$ and $q = 1 - p$, $q \to 1$ as $t \to \infty$, and therefore, all of the curves in Figure 2.1 asymptotically approach unity after a sufficiently long period of time.

[3] This book uses functional block diagrams to describe the functional relationships between system elements that are capable of independent failure. Functional block diagrams are similar to reliability block diagrams but do not strictly adhere to the same conventions. Also, functional block diagrams often require an accompanying explanation to unambiguously describe the characteristics of the system.

a given element may be inoperative owing to the failure of other elements on which it depends, but it still may be capable of failure independently of the other elements in the system. In the following discussion, each element of the overall system is represented as a block, with the appropriate inputs and outputs representing its relationship to the remainder of the system. Each block contains an element p_i that is assumed to have a failure rate λ_i.

Figure 2.2 represents a single component, p_1, of a system that in turn could be part of another block diagram depicting a larger system. This component block has a single input and a single output, but in the more general case, the block might have multiple inputs and multiple outputs. By convention, inputs are generally assumed to enter either from the left side or from the top side of the box, and outputs exit either from the right side or from the bottom side.

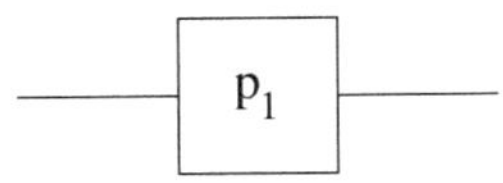

Fig. 2.2 Single-component block diagram

An overall system, of course, is made up of multiple blocks or elements. In general, these elements or groups of elements are arranged either in series or in parallel.

2.3 Elements in Series

The simplest system arrangement involves two elements, each of which has an independent failure mode, that are arranged so that the input of one block is dependent on the output of the previous block. As a result, this system (or portion of a larger system) requires that both blocks be operational. Such an arrangement is termed *elements in series*.

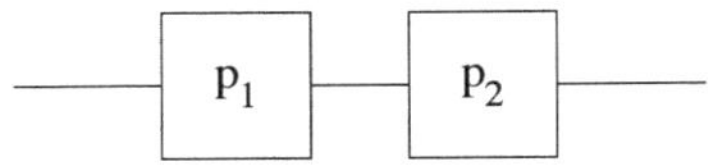

Fig. 2.3 Block diagram for elements in series

The reliability of the system made up of elements p_1 and p_2, as shown in Figure 2.3, is the probability that both components are operational. If the symbols p_1 and p_2 also represent the respective reliability of elements p_1 and p_2 from Figure 2.3[4] and if R_S represents the total reliability of the system that comprises components p_1 and p_2, then

$$R_S = p_1 p_2 \,.$$

It should be clear that this relationship is easily generalized to a system of n elements arranged in series:

$$R_S = \prod_{i=1}^{n} p_i \,. \tag{2.5}$$

During analysis of complex systems, series elements that do not interact with other system elements should be collapsed into a single equivalent element with an assigned reliability equal to the product of the series reliabilities. If the elements have a constant failure rate, which implies an exponential distribution, then the component failure rates are simply added:

$$\lambda_{series} = \sum_{i=1}^{n} \lambda_i \,. \tag{2.6}$$

Thus, λ_{series} is the failure rate of the *new* single element.

A *Mathematica* function can be defined to represent the combination of elements in series. The success of a system made up of two elements arranged in series depends on both elements being operational:

```
rAND[p1_,p2_]:=p1p2;
```

The function name **rAND** was chosen because it represents the probability that both **p1** *and* **p2** are operational. The continued usefulness of the Boolean function analogy is made apparent in later sections.

In a series system, the overall system reliability is a function of individual element reliabilities and the number of elements in series. This relationship is shown in Figure 2.4 for systems composed of 1 to 20 elements,[5] each having an individual element reliability of 0.99, 0.98 or 0.95. Obviously, if the overall reliability of a system consisting of series elements is to be increased, either the individual element reliabilities must be increased or the number of elements in the system must be decreased.

[4] This book uses a symbol, such as p_i, to represent both the given system element and also the reliability of that element. Although this definition leads to some ambiguity, it should not cause confusion in context.

[5] Note that the curves shown in Figure 2.4 are actually defined only for integral numbers of elements in series.

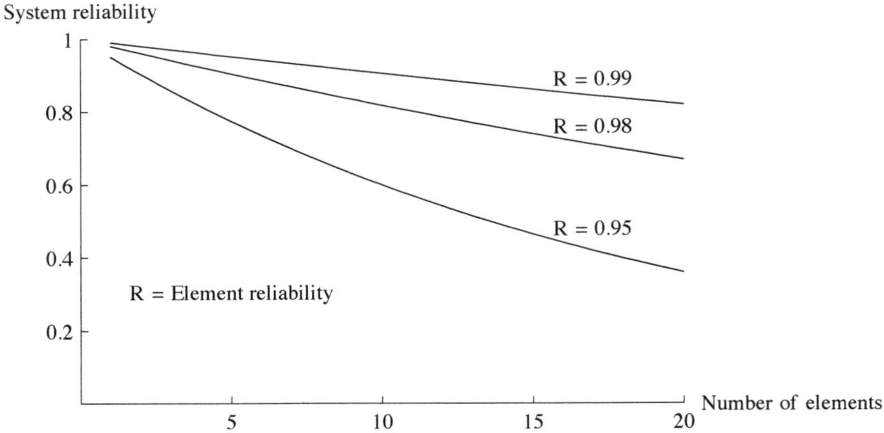

Fig. 2.4 Reliability of series systems

2.4 Elements in Parallel

The second basic arrangement of system components is shown in Figure 2.5. In this arrangement, the system is composed of elements p_1 and p_2 and is operational if either element or both elements are operational.

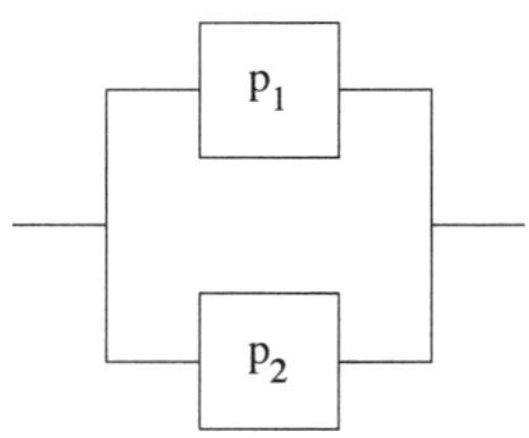

Fig. 2.5 Block diagram for elements in parallel

If p_1 and p_2 represent the reliability of elements p_1 and p_2, respectively, and if R_P is the reliability of the system composed of these two elements, then the system reliability is

$$R_P = 1 - (1 - p_1)(1 - p_2)\,.$$

The general relationship for a system of n components arranged in parallel is

$$R_P = 1 - \prod_{i=1}^{n} (1 - p_i)\,. \tag{2.7}$$

Absent any reasons for showing the redundancy involved in the use of parallel elements, simple parallel arrangements (such as that shown in Figure 2.5) should be collapsed into a single equivalent element with an assigned reliability R_P. This same principle applies to series elements.

The following *Mathematica* function, which represents the combination of elements in parallel, is based on Equation (2.7). The success of a system made up of two elements arranged in parallel depends on either element or both elements being operational:

```
rOR[p1_,p2_]:=1 – (1 – p1)(1 – p2);
```

The Boolean function analogy is employed again by using the name **rOR** for the function.

Figure 2.6 shows the effect of an increasing number of parallel elements on system reliability.[6] The redundant elements have reliabilities of 0.95, 0.9 or 0.8. The reliability of a system with components arranged in parallel increases rapidly with an increasing number of elements, or levels of redundancy, and asymptotically approaches unity with increasing n.

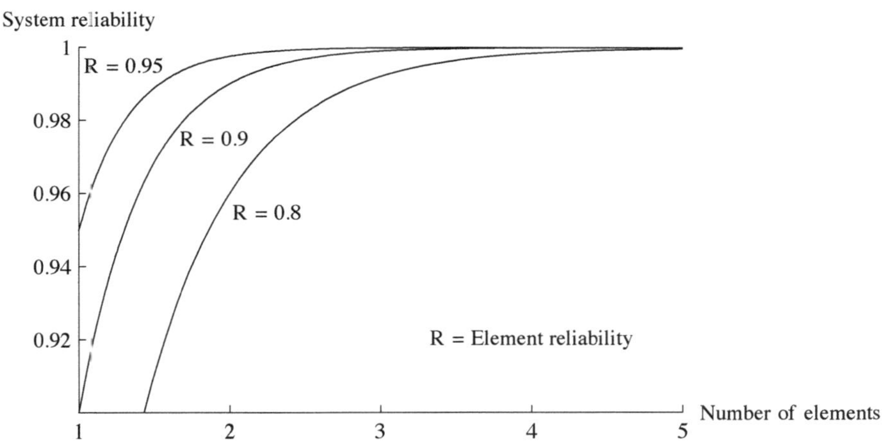

Fig. 2.6 Reliability of parallel systems

2.5 Combined Series/Parallel Systems

Figure 2.7 depicts a system comprising seven elements arranged in series and in parallel. If the elements p_i in Figure 2.7 also have individual element reliabilities

[6] Again, these curves are defined only for integral values of n.

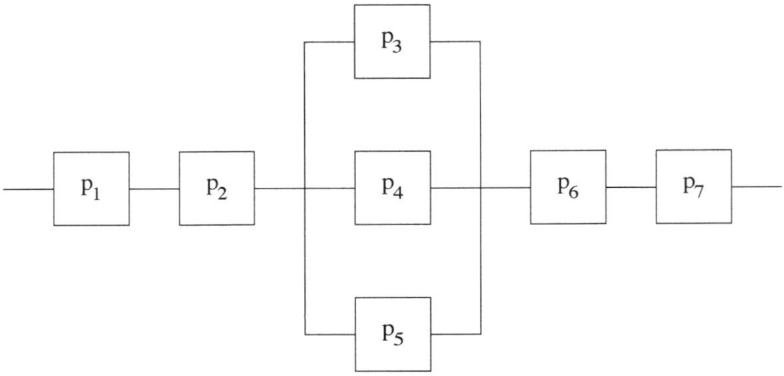

Fig. 2.7 System with elements in series and parallel

p_i, then the overall system reliability R_{sys} can be determined using the *Mathematica* functions defined above:

```
p12 = rAND[p1, p2]
```
$p1p2$

```
p345 = rOR[p3, rOR[p4, p5]]
```
$1 - (1 - p3)(1 - p4)(1 - p5)$

```
p67 = rAND[p6, p7]
```
$p6p7$

```
Rsys = rAND[p12, rAND[p345, p67]]
```
$p1p2(1 - (1 - p3)(1 - p4)(1 - p5))p6p7$

The same result is obtained with a single deeply nested expression:

```
Rsys = rAND[rAND[p1, p2], rAND[rOR[p3, rOR[p4, p5]], rAND[p6, p7]]]
```
$p1p2(1 - (1 - p3)(1 - p4)(1 - p5))p6p7$

This simplification technique can be used to reduce a system of multiple elements, arranged in series and parallel, to an equivalent single-block system as long as all of the elements are simply interconnected. Most real-world complex systems, however, are *not* simply interconnected.

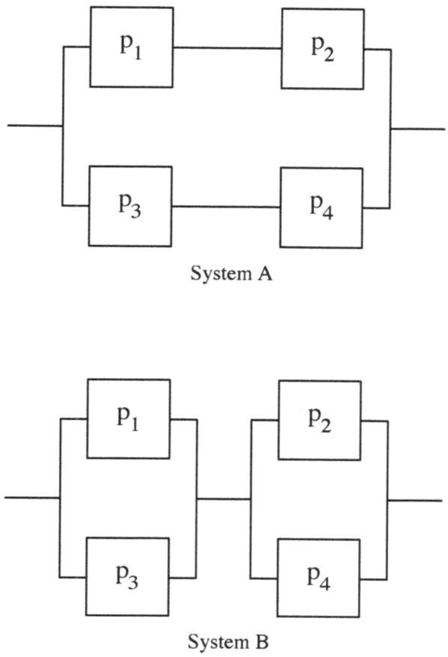

Fig. 2.8 Parallel system arrangements—high-level (System A) and low-level redundancy (System B)

2.6 Parallel System Arrangements

Consider the two alternative parallel system arrangements shown in Figure 2.8. Both of these systems are composed of identical components $p_1, \ldots, p_4$, but the system reliabilities are not equal. For System A to be operational, both p_1 and p_2 or both p_3 and p_4 must be operational. By contrast, System B is operational if at least elements p_1 and p_2, p_1 and p_4, p_3 and p_2, or p_3 and p_4 are operational. System A demonstrates high-level redundancy; System B demonstrates low-level redundancy. The reliability of the two systems can be computed as shown below.

The reliability for System A (high-level redundancy) is

```
rSysA = rOR[rAND[p1, p2], rAND[p3, p4]] // Expand
```
$p1p2 + p3p4 - p1p2p3p4$

and for System B (low-level redundancy) is

```
rSysB = rAND[rOR[p1, p3], rOR[p2, p4]] // Expand
```
$p1p2 - p2p3 - p1p2p3 + p1p4 - p1p2p4 + p3p4 - p1p3p4 - p2p3p4 +$
$p1p2p3p4$

Clearly, the two systems do not have equivalent reliability. If numerical reliability values are assigned to each of the four elements, the difference between the total system reliabilities can be calculated:

```
p1 = .95; p2 = .95; p3 = .9; p4 = .9;
```

```
rSysA
```
0.981475

```
rSysB
```
0.990025

Frequently, however, it is more instructive to look at the system unreliability (that is, the probability of failure) when comparing system alternatives.

```
qSysA = 1 - rSysA
```
0.018525

```
qSysB = 1 - rSysB
```
0.009975

```
qSysA/qSysB
```
1.85714

Note that for these specific component reliabilities, System A is 1.86 times more likely to fail than is System B. In general, the low-level redundancy of System B has greater reliability than the high-level redundancy of System A.

Consider versions of System A and System B in which the components are identical and, consequently, each component has the same reliability. Figure 2.9 shows the ratio of the reliability of System B to that of System A as the element reliabilities are varied over the interval $[0 + \epsilon, 1]$. These are general results; systems with low-level redundancy outperform those with high-level redundancy by as much as a factor of two (in the case of low component reliability). Nevertheless, it should be noted that although the components that compose System A and System B may be identical, System B will have an additional level of complexity for most real-world systems. In most cases, to manage the system redundancy, the configuration of System B requires additional switching logic that would not be required for System A. A basic tenet of system design for reliability, however, is that low-level redundancy outperforms high-level redundancy.

2.7 Redundancy and System Reliability

As shown in the previous section, redundant systems are more reliable than single-strand or series systems. Each additional level of redundancy reduces the likelihood of system failure. Consider the following simplex, duplex and quadruplex systems,

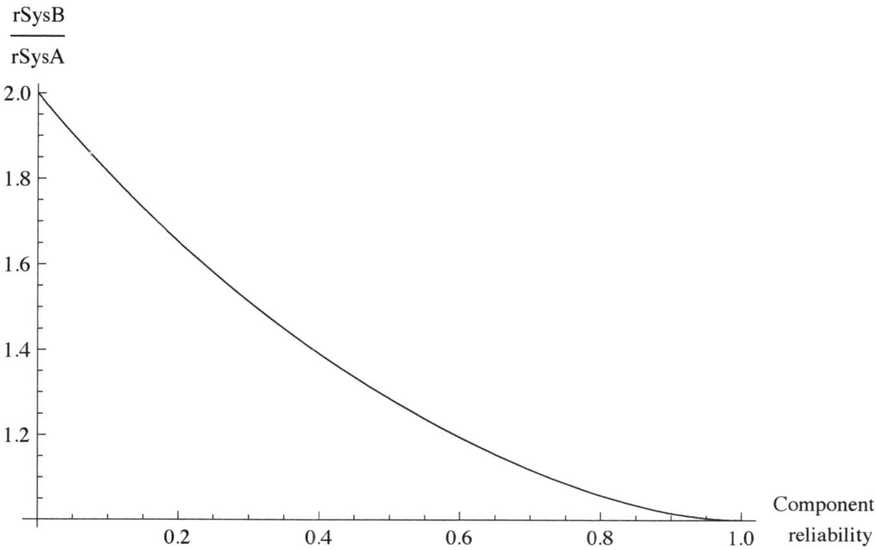

Fig. 2.9 Ratio of low-level to high-level redundant system reliabilities

i.e. systems with one, two, three and four levels of redundancy:

```
rSys1 = p1
```
$p1$

```
rSys2 = rOR[p1, p2]
```
$1-(1-p1)(1-p2)$

```
rSys3 = rOR[p1, rOR[p2, p3]]
```
$1-(1-p1)(1-p2)(1-p3)$

```
rSys4 = rOR[p1, rOR[p2, rOR[p3, p4]]]
```
$1-(1-p1)(1-p2)(1-p3)(1-p4)$

Figure 2.10 shows the probability of failure for each of the systems, given that the reliability of the elements is 0.9 ($p1 = p2 = p3 = p4 = 0.9$). For each additional level of redundancy, the overall reliability of the system increases by an order of magnitude. In this example, the simplex system is one thousand times more likely to fail than is the quadruplex system. The combination of redundancy and high component reliability can yield very low probabilities of system failure.

Figure 2.11 illustrates the relationship between component reliability and the level of redundancy. Note that even with low-reliability elements ($p_i = 0.5$), a comparatively high system reliability (approximately 0.94) can be achieved with four

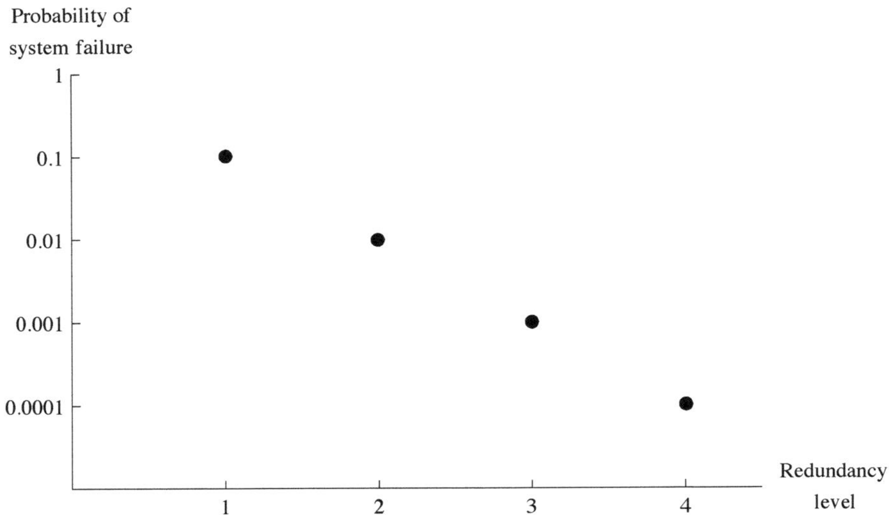

Fig. 2.10 Probability of system failure as a function of redundancy level

elements in parallel. Also note that the benefit of additional parallel elements diminishes rapidly as the redundancy level increases beyond three or four.

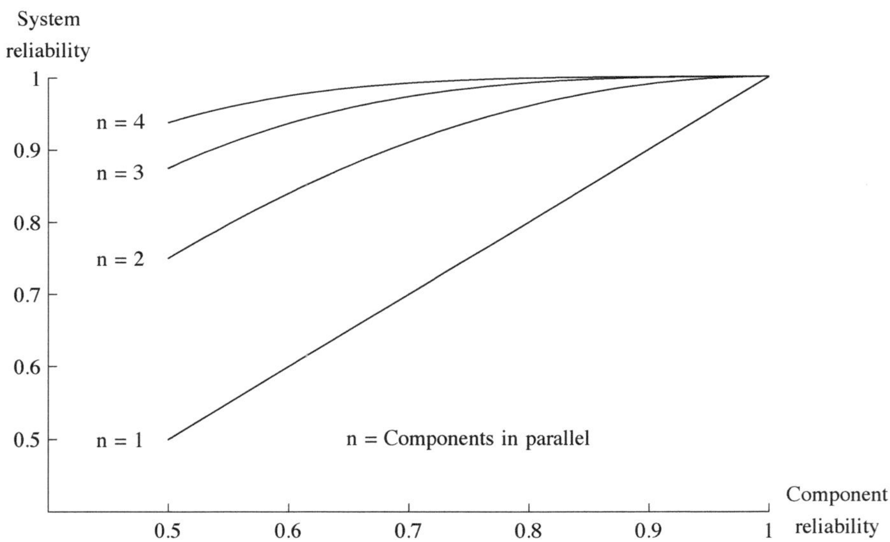

Fig. 2.11 Redundant system reliability as a function of component reliability

The results shown in Figures 2.10 and 2.11 are for redundant components with constant reliabilities; for real systems, understanding the system reliability as a function of time may be more useful. As previously discussed, components with a constant failure rate λ have a probability of failure function $p_i = e^{-\lambda_i t}$. Figure 2.12 shows the probability of failure for simplex, duplex and quadruplex systems with identical components for which $\lambda = 1000$ fpmh.

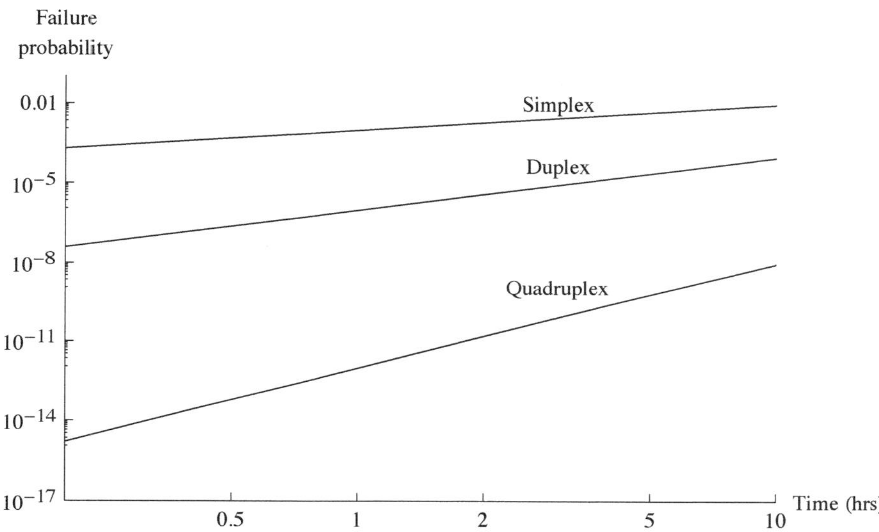

Fig. 2.12 Probability of failure for redundant systems ($\lambda = 1000$ fpmh)

When shown in log-log scale, each of the system failure probability curves is a straight line with a slope, measured in decades per decade (dpd), that is nearly equal to the redundancy of the system. For redundant systems with components having an exponential failure distribution, the slope of the probability of failure curve plotted on log-log axes is a measure of the system's redundancy level. In this book, this slope is referred to as the *equivalent redundancy level* (ERL) of the system.

Over the interval $[0, 10]$, the system failure probability curves shown in Figure 2.12 are very nearly straight lines. For large mission times, the system failure probability curves asymptotically approach unity over the interval $[0, 10^4]$, as shown in Figure 2.13. Once the curvature starts to appear, however, it can be argued that the system is past its useful lifetime.

Figure 2.14 illustrates the effect of increasing the component failure rates from 1000 fpmh (solid curves) to 2000 fpmh (dashed curves). The slope of the failure probability curve indicates the effective level of redundancy, and the vertical displacement is a function of the redundant component failure rate. Note that the vertical displacement is four times greater for the quadruplex system than for the simplex system. These relationships are general, and as a result, depicting the probability of

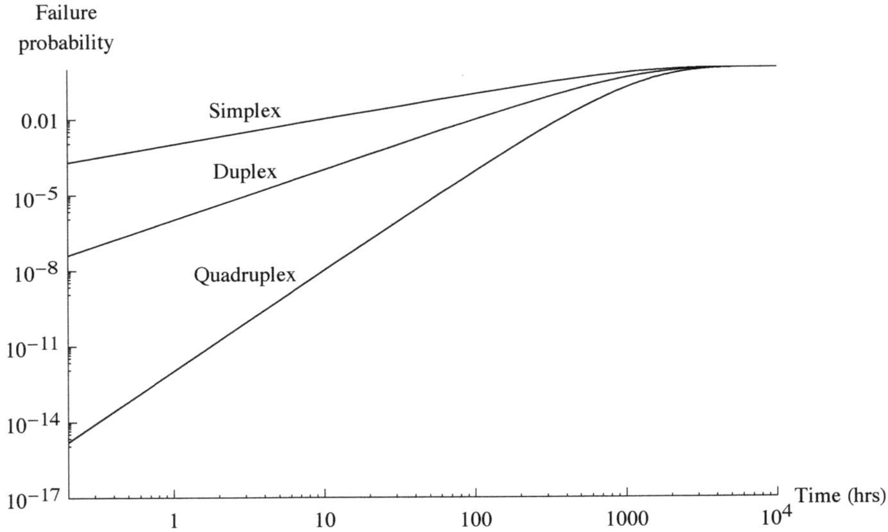

Fig. 2.13 Probability of failure for redundant systems for large mission times (λ = 1000 fpmh)

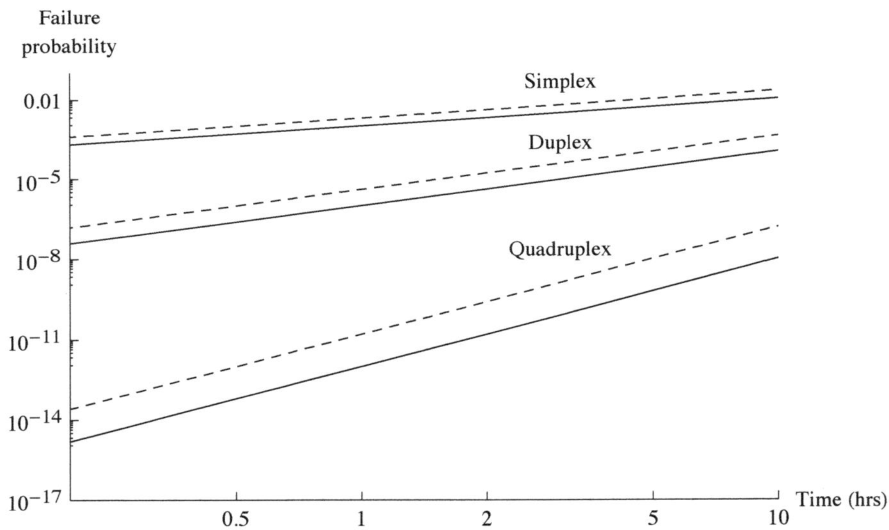

Fig. 2.14 Probability of failure for redundant systems, λ = 1000 fpmh (solid lines) and 2000 fpmh (dashed lines)

system failure on a log-log scale is helpful. Furthermore, the relationships will generally hold even if the redundant elements are composed of multiple components that constitute a full channel in a multichannel system.

For a system with perfect fault coverage, the slope of the probability of failure curve, when plotted on a log-log scale, approximates the level of system redundancy; that is, a quadruplex system has a slope of approximately 4 dpd, a triplex system has a slope of approximately 3 dpd and so on. Again, for this reason, the slope of a system's probability of failure curve is referred to as its *equivalent redundancy level*, or ERL.

These results illustrate two general effects: the redundancy level determines the slope of a redundant system's probability of failure curve, and the redundant component reliability shifts the curve vertically.[7]

2.8 *k*-out-of-*n*:G Systems

The redundant systems discussed in Sections 2.4 and 2.7 are examples of at least 1-out-of-n:G systems, where the system is operational (G → good) if at least one of the n redundant components is operational. The series systems discussed in Section 2.3 are n-out-of-n:G systems, where all n of the components must be operational for the system to be operational. Both parallel and series systems are examples of the more general k-out-of-n:G system structure in which at least k of the n system components must be operational for the system to function.

In addition to determining *at least* k-out-of-n:G system reliability, there are circumstances for which determining the reliability of an *exactly* k-out-of-n:G system is also useful, and both are presented in the following sections.

2.8.1 At Least k-out-of-n:G Systems

If a system consists of identical components, then the components are independent and also have identical failure distributions; the components are said to be *independent and identically distributed* (i.i.d.). For an i.i.d. k-out-of-n:G system with perfect fault coverage, the system reliability can be readily computed using the following equation:

[7] These general results are applicable to redundant systems that are *not* subject to imperfect fault coverage. In subsequent chapters, the effect of imperfect fault coverage is examined in some detail. The probability of failure curves plot on a log-log scale as straight lines as long as λt is within an effective range. Obviously, as $\lambda t \rightarrow \infty$, the probability of failure approaches unity. For highly reliable redundant elements, however, the probability of failure curve continues to be a straight line for reasonable mission times.

$$R(k,n) = \sum_{i=k}^{n} \binom{n}{i} p^i (1-p)^{n-i} \tag{2.8}$$
$$= \sum_{i=k}^{n} \binom{n}{i} p^i q^{n-i} \,.$$

In Equation (2.8), $q = 1 - p$. This equation uses a widely known function (see, for instance, page 21 in Barlow and Proschan [1]).

For the general case with non-identical components, computing the system reliability is somewhat more complex. Let $\mathbf{p} = \{p_1, \ldots, p_n\}$ be a vector of component reliabilities (which are not necessarily i.i.d.), and likewise, let $\mathbf{q} = \{1-p_1, \ldots, 1-p_n\}$ be a vector of component unreliabilities. Then define the following functions:

$\mathbf{pT}(i, \mathbf{p})$ Set of products of the k-subsets of $\mathbf{p}$ with exactly i elements
$\mathbf{qT}(i, \mathbf{q})$ Set of products of the k-subsets of $\mathbf{q}$ with exactly $(n - i)$ elements

A k-subset is a subset with exactly k elements.[8] Note that the k-subsets must be generated in lexicographic order. The reliability of a non-i.i.d. k-out-of-n:G system is given by

$$R(k, n, \mathbf{p}) = \sum_{i=k}^{n} \sum_{j=1}^{\binom{n}{i}} \mathbf{qT}(i, \mathbf{p})_j \mathbf{pT}(i, \mathbf{p})_{\binom{n}{i}-j+1} \,. \tag{2.9}$$

For example, given $\mathbf{p} = \{p_1, p_2, p_3\}$ so that $n = |\mathbf{p}| = 3$, the required $\mathbf{pT}$ and $\mathbf{qT}$ sets are

$\mathbf{pT}(1, \mathbf{p}) = \{p_1, p_2, p_3\}$
$\mathbf{pT}(2, \mathbf{p}) = \{p_1 p_2, p_1 p_3, p_2 p_3\}$
$\mathbf{pT}(3, \mathbf{p}) = \{p_1 p_2 p_3\}$

$\mathbf{qT}(1, \mathbf{p}) = \{(1 - p_1)(1 - p_2), (1 - p_1)(1 - p_3), (1 - p_2)(1 - p_3)\}$
$\mathbf{qT}(2, \mathbf{p}) = \{(1 - p_1), (1 - p_2), (1 - p_3)\}$
$\mathbf{qT}(3, \mathbf{p}) = 1$,

leading to the following:

$$\begin{aligned} R(1, 3, \mathbf{p}) &= p_1(1 - p_2)(1 - p_3) + (1 - p_1)p_2(1 - p_3) \\ &\quad + p_1 p_2(1 - p_3) + (1 - p_1)(1 - p_2)p_3 \\ &\quad + p_1(1 - p_2)p_3 + (1 - p_1)p_2 p_3 + p_1 p_2 p_3 \\ &= p_1 + p_2 - p_1 p_2 + p_3 - p_1 p_3 - p_2 p_3 \\ &\quad + p_1 p_2 p_3 \,. \end{aligned} \tag{2.10}$$

[8] Including the set itself and the empty set, a set of n elements has 2^n subsets. A *k-subset* is a subset with exactly k elements [4].

Additional detail on the derivation and use of Equation (2.9) is given in [3]. An algorithm and computer code for the generation of k-subsets in lexicographic order is given in [2]. In later chapters, Equation (2.9) is modified to include the calculation of general k-out-of-n:G systems subject to imperfect fault coverage.

2.8.2 *Exactly k-out-of-n:G Systems*

The probability that an *at least* k-out-of-n:G system is operational is simply the sum from k to n of the probability that *exactly* k out of n components are operational. This leads to straightforward functions for determining exactly k-out-of-n:G system reliability.

For a general exactly k-out-of-n:G system with components that are not necessarily i.i.d.,

$$R_e(k, n, \mathbf{p}) = \sum_{j=1}^{\binom{n}{k}} \mathbf{qT}(k, \mathbf{p})_j \mathbf{pT}(k, \mathbf{p})_{\binom{n}{k}-j+1} \; . \tag{2.11}$$

For an exactly k-out-of-n:G system with i.i.d. components, Equation (2.11) can be simplified to the following well-known expression [1]:

$$\begin{aligned} R_e(k, n) &= \binom{n}{k} p^k (1-p)^{n-k} \\ &= \binom{n}{k} p^k q^{n-k} \; . \end{aligned} \tag{2.12}$$

2.8.3 Mathematica *k-out-of-n:G Reliability*

2.8.3.1 *Mathematica* i.i.d. k-out-of-n:G Reliability

The combinatorial functions presented in this section can be easily formulated as *Mathematica* functions. For Equation (2.8) in the case of i.i.d. components and at least k-out-of-n:G reliability,

```
Ral[k_,n_,p_]:=Module[{i, q = (1 - p)},
∑_{i=k}^{n} Binomial[n, i] p^i q^(n-i) //Expand];
```

For Equation (2.12) in the case of i.i.d. components and exactly k-out-of-n:G reliability,

```
Rex[k_,n_,p_]:=Module[{q = (1 − p)},
Binomial[n, k]p^k q^(n−k)//Expand];
```

The **Ral** function returns a fully expanded polynomial representing the i.i.d. at least k-out-of-n:G system reliability, and the **Rex** function returns a fully expanded polynomial representing the i.i.d. exactly k-out-of-n:G system reliability. For example, the reliabilities for i.i.d. at least 3-out-of-7:G and exactly 3-out-of-7:G systems are

```
Ral[3,7,p]
```
$35p^3 - 105p^4 + 126p^5 - 70p^6 + 15p^7$

and

```
Rex[3,7,p]
```
$35p^3 - 140p^4 + 210p^5 - 140p^6 + 35p^7$

2.8.3.2 *Mathematica* Non i.i.d. k-out-of-n:G Reliability

The *Mathematica* functions for non i.i.d. k-out-of-n:G reliability are somewhat more complicated but are actually straightforward implementations of Equations (2.9) and (2.11).

Define the required **pT** and **qT** functions (the *Mathematica* function **KSubsets** requires the **Combinatorica** package):

```
Needs["Combinatorica`"];

pT[i_Integer,p_List]:=Module[{},
Apply[Times,KSubsets[p,i],{1}]];

qT[i_Integer,p_List]:=
Module[{j,n = Length[p],q},q = Table[1 − p[[j]],{j,n}];
Apply[Times,KSubsets[q,n − i],{1}]];
```

Define the following functions:

```
Ral[k_,p_List]:=Module[{i,j,n = Length[p]},
```
$\sum_{i=k}^{n}\sum_{j=1}^{\text{Binomial}[n,i]}$ `qT[i,p][[j]]pT[i,p][[Binomial[n,i] − j + 1]]//Expand];`

and

```
Rex[k_, p_List] := Module[{j, n = Length[p]},
  Sum[qT[k, p][[j]] pT[k, p][[Binomial[n, k] - j + 1]], {j, 1, Binomial[n, k]}] // Expand];
```

Consider, for example, at least 1-out-of-4:G reliability and exactly 1-out-of-4:G reliability, both with non i.i.d. components:

```
Ral[1, {p1, p2, p3, p4}]
p1 + p2 − p1p2 + p3 − p1p3 − p2p3 + p1p2p3 + p4 −
p1p4 − p2p4 + p1p2p4 − p3p4 + p1p3p4 + p2p3p4 − p1p2p3p4
```

and

```
Rex[1, {p1, p2, p3, p4}]
p1 + p2 − 2p1p2 + p3 − 2p1p3 − 2p2p3 + 3p1p2p3 + p4 − 2p1p4 −
2p2p4 + 3p1p2p4 − 2p3p4 + 3p1p3p4 + 3p2p3p4 − 4p1p2p3p4
```

The *Mathematica* substitution **pn** → **p** permits the computation of the same i.i.d. 3-out-of-7:G reliability as the one calculated above:

```
ToP = {p1 → p, p2 → p, p3 → p, p4 → p, p5 → p, p6 → p, p7 → p};
```

```
Ral[3, {p1, p2, p3, p4, p5, p6, p7}] /. ToP
```

$35p^3 - 105p^4 + 126p^5 - 70p^6 + 15p^7$

and

```
Rex[3, {p1, p2, p3, p4, p5, p6, p7}] /. ToP
```

$35p^3 - 140p^4 + 210p^5 - 140p^6 + 35p^7$

These results are identical to those obtained above using the i.i.d. variations of the **Ral** and **Rex** functions defined in Section 2.8.3.1. Note that even though the non i.i.d. **Ral** and **Rex** functions defined here have the same names as the previously defined functions, *Mathematica* is able to distinguish between them because they have different arguments. Consequently, the implementations do not conflict.

This section has presented combinatorial functions for the calculation of *k*-out-of-*n*:G system reliability along with implementation as *Mathematica* functions. These combinatorial expressions have a computational complexity of $O(\sum_{i=k}^{n} \binom{n}{i})$, which may be unacceptable for large n values ($n \gtrsim 10$). Other algorithms (table-based codes), which are presented in later chapters, yield identical results but have a complexity of $O(n \cdot (n-k))$. This order of complexity permits efficient calculation, even for large values of n.

References

1. Barlow RE, Proschan F (1975) Statistical Theory of Reliability and Life Testing: Probability Models. Holt, Reinhart and Winston, New York
2. Buckles BP, Lybanon M (1977) Algorithm 515: Generation of a vector from the lexicographical index. ACM Trans Math Softw 3:180–182
3. Myers AF (2007) k-out-of-n:G System Reliability With Imperfect Fault Coverage. IEEE Trans on Reliab 56:464–473
4. Weisstein EW (1999) CRC Concise Encyclopedia of Mathematics. CRC Press, USA

Chapter 3
Complex System Reliability

"Everything should be made as simple as possible,
but no simpler."
—Albert Einstein

Abstract All of the systems studied in Chapter 2 are simply interconnected; most real-world multichannel systems, however, are not simply interconnected. As a consequence, they cannot be simplified to a single equivalent element or block through some combination of series and parallel reductions. Systems that do not have simple interconnections are called *complex*, and it is the assessment of complex systems that makes the general reliability problem NP-complete.

3.1 Systems with Complex Interconnections

Consider the system depicted in Figure 3.1, which, in texts that cover system reliability analysis, has frequently been used as an example of a system with complex interconnections. This system cannot be analyzed by way of repeated simplifications using series, parallel or k-out-of-n:G techniques. The reason for this complexity is that p_4 and p_5 are both partially dependent on the output of p_2.

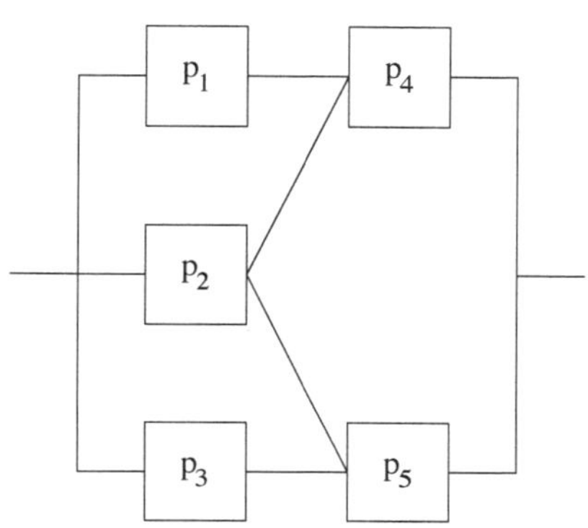

Fig. 3.1 Simple system with complex interconnections

For a naive first attempt to analyze this system, the **rAND** and **rOR** functions described in Sections 2.3 and 2.4 can be used. Recall that each of the elements $p_1, \ldots, p_5$ of **p** are binary random variables (also called *Bernoulli variables*). The system is operational if the output of p_4, the output of p_5 or both are operational. By determining the output of both p_4 and p_5, an attempt can be made at computing the reliability of the system:

```
p4out = rOR[rAND[p2,p4],rAND[p1,p4]]
```
$1-(1-\text{p1p4})(1-\text{p2p4})$

```
p5out = rOR[rAND[p2,p5],rAND[p3,p5]]
```
$1-(1-\text{p2p5})(1-\text{p3p5})$

```
rSysIncorrect = rOR[p4out,p5out]
```
$1-(1-\text{p1p4})(1-\text{p2p4})(1-\text{p2p5})(1-\text{p3p5})$

```
rSysIncorrect//Expand
```
$\text{p1p4}+\text{p2p4}-\text{p1p2p4}^2+\text{p2p5}+\text{p3p5}-\text{p1p2p4p5}-\text{p2}^2\text{p4p5}-$
$\text{p1p3p4p5}-\text{p2p3p4p5}+\text{p1p2}^2\text{p4}^2\text{p5}+\text{p1p2p3p4}^2\text{p5}-$
$\text{p2p3p5}^2+\text{p1p2p3p4p5}^2+\text{p2}^2\text{p3p4p5}^2-\text{p1p2}^2\text{p3p4}^2\text{p5}^2$

The polynomial **rSysIncorrect** is distinctly different from those encountered in Chapter 2, where none of the literals in the reliability polynomials were raised to a power greater than unity. Because the elements p_i are Bernoulli variables that only exist in state 1 or state 0 and because both $1^n = 1$ and $0^n = 0$, the presence of literals raised to a power greater than unity is not meaningful. The expression above contains multiple instances in which literals are raised to the second power, and if the expression was used to compute system reliability, the result would be incorrect.

The next section presents a technique that uses a sum-over-states (or truth-table) approach for obtaining the correct reliability of a system with complex interconnections. A sum-over-states approach can, at least in principle, obtain correct results for any system.

3.2 Sum over States and Truth Tables

One approach to computing the correct reliability value for the system in Figure 3.1 is summation of the probabilities of all the states for which the system is operational. A table listing the probability of each possible state and its consequences for a system is frequently referred to as a *truth table*, and the use of the information contained in a truth table to determine system reliability is called a *sum over states* (SOS). The system shown in Figure 3.1 has five elements, $p_1, \ldots, p_5$, and it therefore has $2^5 = 32$ possible states. Only a portion of these states yield an operational system. The possible states are summarized in the form of a truth table, shown in Table 3.1, with each element being assigned either an operational state, p_i, or a failed state, q_i. The P(state) column lists the probabilities of each of the n system states. Since the table covers all of the possible combinations, the sum of all of the state probabilities must be unity. The state probabilities resulting in an operational system are included in the column labeled P(OP). The sum of the operational system state probabilities included in this column is the reliability of the system.

Table 3.1 Possible states for the system in Figure 3.1

State	p_1	p_2	p_3	p_4	p_5	P(state)	P(OP)
1	q_1	q_2	q_3	q_4	q_5	$q_1q_2q_3q_4q_5$	
2	p_1	q_2	q_3	q_4	q_5	$p_1q_2q_3q_4q_5$	
3	q_1	p_2	q_3	q_4	q_5	$q_1p_2q_3q_4q_5$	
4	p_1	p_2	q_3	q_4	q_5	$p_1p_2q_3q_4q_5$	
5	q_1	q_2	p_3	q_4	q_5	$q_1q_2p_3q_4q_5$	
6	p_1	q_2	p_3	q_4	q_5	$p_1q_2p_3q_4q_5$	
7	q_1	p_2	p_3	q_4	q_5	$q_1p_2p_3q_4q_5$	
8	p_1	p_2	p_3	q_4	q_5	$p_1p_2p_3q_4q_5$	
9	q_1	q_2	q_3	p_4	q_5	$q_1q_2q_3p_4q_5$	
10	p_1	q_2	q_3	p_4	q_5	$p_1q_2q_3p_4q_5$	$p_1q_2q_3p_4q_5$
11	q_1	p_2	q_3	p_4	q_5	$q_1p_2q_3p_4q_5$	$q_1p_2q_3p_4q_5$
12	p_1	p_2	q_3	p_4	q_5	$p_1p_2q_3p_4q_5$	$p_1p_2q_3p_4q_5$
13	q_1	q_2	p_3	p_4	q_5	$q_1q_2p_3p_4q_5$	
14	p_1	q_2	p_3	p_4	q_5	$p_1q_2p_3p_4q_5$	$p_1q_2p_3p_4q_5$
15	q_1	p_2	p_3	p_4	q_5	$q_1p_2p_3p_4q_5$	$q_1p_2p_3p_4q_5$
16	p_1	p_2	p_3	p_4	q_5	$p_1p_2p_3p_4q_5$	$p_1p_2p_3p_4q_5$
17	q_1	q_2	q_3	q_4	p_5	$q_1q_2q_3q_4p_5$	
18	p_1	q_2	q_3	q_4	p_5	$p_1q_2q_3q_4p_5$	
19	q_1	p_2	q_3	q_4	p_5	$q_1p_2q_3q_4p_5$	$q_1p_2q_3q_4p_5$
20	p_1	p_2	q_3	q_4	p_5	$p_1p_2q_3q_4p_5$	$p_1p_2q_3q_4p_5$
21	q_1	q_2	p_3	q_4	p_5	$q_1q_2p_3q_4p_5$	$q_1q_2p_3q_4p_5$
22	p_1	q_2	p_3	q_4	p_5	$p_1q_2p_3q_4p_5$	$p_1q_2p_3q_4p_5$
23	q_1	p_2	p_3	q_4	p_5	$q_1p_2p_3q_4p_5$	$q_1p_2p_3q_4p_5$
24	p_1	p_2	p_3	q_4	p_5	$p_1p_2p_3q_4p_5$	$p_1p_2p_3q_4p_5$
25	q_1	q_2	q_3	p_4	p_5	$q_1q_2q_3p_4p_5$	
26	p_1	q_2	q_3	p_4	p_5	$p_1q_2q_3p_4p_5$	$p_1q_2q_3p_4p_5$
27	q_1	p_2	q_3	p_4	p_5	$q_1p_2q_3p_4p_5$	$q_1p_2q_3p_4p_5$
28	p_1	p_2	q_3	p_4	p_5	$p_1p_2q_3p_4p_5$	$p_1p_2q_3p_4p_5$
29	q_1	q_2	p_3	p_4	p_5	$q_1q_2p_3p_4p_5$	$q_1q_2p_3p_4p_5$
30	p_1	q_2	p_3	p_4	p_5	$p_1q_2p_3p_4p_5$	$p_1q_2p_3p_4p_5$
31	q_1	p_2	p_3	p_4	p_5	$q_1p_2p_3p_4p_5$	$q_1p_2p_3p_4p_5$
32	p_1	p_2	p_3	p_4	p_5	$p_1p_2p_3p_4p_5$	$p_1p_2p_3p_4p_5$

```
rSys = p1p4q2q3q5 + p2p4q1q3q5 + p1p2p4q3q5 +
p1p3p4q2q5 + p2p3p4q1q5 + p1p2p3p4q5 +
p2p5q1q3q4 + p1p2p5q3q4 + p3p5q1q2q4 +
p1p3p5q2q4 + p2p3p5q1q4 + p1p2p3p5q4 +
p1p4p5q2q3 + p2p4p5q1q3 + p1p2p4p5q3 +
```

```
p3p4p5q1q2 + p1p3p4p5q2 + p2p3p4p5q1 +
p1p2p3p4p5;
```

The following substitution converts the sum from a polynomial in q_i and p_i to one solely in p_i.

```
rSys = rSys/.q1 → (1 − p1)/.q2 → (1 − p2)/.q3 → (1 − p3)/.
q4 → (1 − p4)/.q5 → (1 − p5)//Expand
p1p4 + p2p4 − p1p2p4 + p2p5 + p3p5 −
p2p3p5 − p2p4p5 − p1p3p4p5 + p1p2p3p4p5
```

The polynomial computed above, **rSys,** is the correct reliability for the system shown in Figure 3.1; it is clearly not equal to the **rSysIncorrect** result obtained previously. The technique of summing all of the operational state probabilities always produces a correct value for the reliability of a system.

The system unreliability could have been calculated by summing all of the non-operational state probabilities, which are those P(state) entries that are not included in the P(OP) column. The system reliability is then unity minus the resulting sum. For this particular case, the calculation of the system unreliability involves summing fewer terms (13 terms) than does the calculation of the system reliability (19 terms). This is not generally the case, however.

```
qSys = q1q2q3q4q5 + p1q2q3q4q5 + p2q1q3q4q5 +
p1p2q3q4q5 + p3q1q2q4q5 + p1p3q2q4q5 +
p2p3q1q4q5 + p1p2p3q4q5 + p4q1q2q3q5 +
p3p4q1q2q5 + p5q1q2q3q4 + p1p5q2q3q4 +
p4p5q1q2q3;
```

```
qSys = qSys/.q1 → (1 − p1)/.q2 → (1 − p2)/.q3 → (1 − p3)/.
q4 → (1 − p4)/.q5 → (1 − p5)//Expand;
```

The above expression produces the following result, which is identical to the **rSys** result above:

```
1 − qSys
p1p4 + p2p4 − p1p2p4 + p2p5 + p3p5 −
p2p3p5 − p2p4p5 − p1p3p4p5 + p1p2p3p4p5
```

The two expressions sum to unity:

```
rSys + qSys
1
```

The result obtained above, **1 − qSys,** agrees with the previously obtained **rSys** result. Also notice that the sum of **rSys** and **qSys** is unity; this is a good check of the procedure, since the two probabilities span the entire set of possibilities and therefore must sum to unity.

Clearly, SOS faces serious shortcomings when used to assess systems with large numbers of components. For the five-component system in the example above, it was necessary to determine $2^5 = 32$ state probabilities and to identify which states constitute success paths. For a "real-world" system with even a modest number of components, the number of states that must be examined is too large to be practical, even by computer standards. For instance, a system with 50 components would involve a staggering $2^{50} \approx 1.126 \times 10^{15}$ states.

3.3 Bernoulli State Variables (BSV)

The *Mathematica* functions **rAND** and **rOR,** which are defined in Sections 2.3 and 2.4, were shown to produce incorrect results for systems with complex interconnections. Nevertheless, by noting that the variables that define the structure of a system are Bernoulli random variables (meaning that they only hold values of 1 or 0), the **rAND** and **rOR** functions can be modified to provide correct results. If p_i is a Bernoulli random variable, then $p_i^n = p_i$, since $1^n = 1$ and $0^n = 0$. This relationship can be implemented using a substitution rule; when this rule is applied to fully expanded polynomials, correct reliability results are obtained.

Define the substitution rule **BernoulliRule**

```
BernoulliRule = x_^n_ → x;
```

Recall from Section 3.1 the previously computed incorrect result, **rSysIncorrect:**

```
rSysIncorrect//Expand
```

$p1p4 + p2p4 - p1p2p4^2 + p2p5 + p3p5 - p1p2p4p5 - p2^2p4p5 - p1p3p4p5 - p2p3p4p5 + p1p2^2p4^2p5 + p1p2p3p4^2p5 - p2p3p5^2 + p1p2p3p4p5^2 + p2^2p3p4p5^2 - p1p2^2p3p4^2p5^2$

To obtain the correct result, the **BernoulliRule** substitution can be applied to the incorrect result:

```
(rSysIncorrect//Expand)/.BernoulliRule
```

$p1p4 + p2p4 - p1p2p4 + p2p5 + p3p5 - p2p3p5 - p2p4p5 - p1p3p4p5 + p1p2p3p4p5$

This application of the **BernoulliRule** substitution provides the correct reliability of the system shown in Figure 3.1 and yields the same result as the **rSys** expression obtained using the sum-over-states approach above.

Incorporating the **BernoulliRule** substitution in a redefinition of the **rAND** and **rOR** functions is more useful and efficient.

```
rAND[p1_,p2_]:=((p1p2)//Expand)/.BernoulliRule;

rOR[p1_,p2_]:=((1 - (1 - p1)(1 - p2))//Expand)/.BernoulliRule;
```

Using the redefined **rAND** and **rOR** functions, the correct solution for the system defined in Section 3.1 can be computed using the same approach as that used in Chapter 2:

```
p4out = rOR[rAND[p2,p4],rAND[p1,p4]]
p1p4 + p2p4 - p1p2p4

p5out = rOR[rAND[p2,p5],rAND[p3,p5]]
p2p5 + p3p5 - p2p3p5

rSys = rOR[p4out,p5out]
p1p4 + p2p4 - p1p2p4 + p2p5 + p3p5 -
p2p3p5 - p2p4p5 - p1p3p4p5 + p1p2p3p4p5
```

This result is in agreement with the solution computed using the sum over the path states from Section 3.2. This approach is called the use of *Bernoulli state variables* (BSV). The next section uses BSV to define operators that replace the **rAND** and **rOR** functions and that provide a concise description of a system reliability problem.

3.4 BernoulliRule and the ⊗ and ⊕ Operators

It is useful to define a set of operators to replace the **rAND** and **rOR** functions. The ⊗ operator replaces the **rAND** function, and the ⊕ replaces the **rOR** function.

> **Note**
> The definition of the ⊕ operator to represent the *or* function is different from the convention sometimes used in electrical engineering, where ⊕ represents the *xor* (exclusive or) function.

Define the ⊗ (**CircleTimes**) and ⊕ (**CirclePlus**) operators:

```
Clear[CircleTimes,CirclePlus];

CircleTimes[x_,x_]:=(x//Expand)/.BernoulliRule;
```

```
CircleTimes[x_,y_]:=((x * y)//Expand)/.BernoulliRule;

CircleTimes[x_,y_,z__]:=CircleTimes[x,CircleTimes[y,z]];

CirclePlus[x_,x_]:=(x//Expand)/.BernoulliRule;

CirclePlus[x_,y_]:=((x + y - x ⊗ y)//Expand)/.BernoulliRule;

CirclePlus[x_,y_,z__]:=CirclePlus[x,CirclePlus[y,z]];

CirclePlus[x_]:=(x//Expand)/.BernoulliRule;
```

Note
In the function definitions above, the third argument, z__, is z followed by two "_" (underscore) characters, which indicates that z can stand for any sequence of one or more expressions. This allows **CirclePlus** and **CircleTimes** to accept any number of arguments.

Using these operators, the reliability of the system depicted in Figure 3.1 can be computed in a straightforward and concise fashion:

```
rSys = (p1 ⊗ p4) ⊕ (p2 ⊗ p4) ⊕ (p2 ⊗ p5) ⊕ (p3 ⊗ p5)
p1p4 + p2p4 - p1p2p4 + p2p5 + p3p5 -
p2p3p5 - p2p4p5 - p1p3p4p5 + p1p2p3p4p5
```

This result is in agreement with the solutions obtained in Sections 3.2 and 3.3.

The **BernoulliRule** substitution can be used to define new versions of the **Ral** and **Rex** functions, which were previously defined in Section 2.8.3. The redefined functions can be used to determine complex system reliability.

The function **RaL,** which is defined below, computes the reliability of at least k-out-of-n:G systems with perfect fault coverage:

```
RaL[k_,p_List]:=Module[{n = Length[p]},
((∑_{i=k}^{n} ∑_{j=1}^{Binomial[n,i]} qT[i,p][[j]]pT[i,p][[Binomial[n,i] - j + 1]]) //Expand)
/.BernoulliRule];
```

Calculation of the **ReX** function yields the reliability of exactly k-out-of-n:G systems with perfect fault coverage:

```
ReX[k_,p_List]:=Module[{n = Length[p]},
((∑_{j=1}^{Binomial[n,k]} qT[k,p][[j]]pT[k,p][[Binomial[n,k] - j + 1]]) //Expand)
/.BernoulliRule];
```

The **Ral** and **Rex** functions that were previously defined in Section 2.8.3 assume that the redundant elements are disjoint,[1] and consequently, these functions do not provide correct results for complex systems. The **RaL** and **ReX** functions defined above, after applying the **BernoulliRule** substitution to the expanded results, yield correct results in the general case, even if the redundant elements are not necessarily disjoint.

3.5 BSV Operations Using ⊗ and ⊕

Mathematica is general in its default handling of variables; consequently, the definition given above for the ⊕ and ⊗ operators supports operation on either vector or scalar operands.

First consider the use of ⊕ and ⊗ on the scalar variables a and b:

```
Clear[a, b];
a ⊗ b
ab
a ⊕ b
a + b − ab
```

If the operations are applied to vectors of equal length, they are applied on a pair-wise basis, element by element:

```
A = {a1, a2, a3, a4};
B = {b1, b2, b3, b4};
A ⊗ B
{a1 b1, a2 b2, a3 b3, a4 b4}
A ⊕ B
{a1 + b1 − a1 b1, a2 + b2 − a2 b2, a3 + b3 − a3 b3, a4 + b4 − a4 b4}
```

The ability to operate on vectors simplifies the task of modeling redundant systems. In the example above, **A** and **B** might represent sets of quad-redundant system components, for instance. In Chapter 5, it will be shown that redundant systems can be modeled in a similar fashion, even though the redundant elements may be represented by complex reliability polynomials rather than simply by individual component reliabilities.

Since the ⊕ and ⊗ operators define operations on Bernoulli random variables, all of the laws and theorems associated with Boolean algebra will hold.

```
Clear[a, b, c];
```

[1] Two sets—redundant components in this case—are disjoint if they have no elements in common.

Commutative law:
$\boldsymbol{a \otimes b == b \otimes a}$
True
$\boldsymbol{a \oplus b == b \oplus a}$
True

Associative law:
$\boldsymbol{a \otimes (b \otimes c) == (a \otimes b) \otimes c}$
True
$\boldsymbol{a \oplus (b \oplus c) == (a \oplus b) \oplus c}$
True

Distributive law:
$\boldsymbol{a \otimes (b \oplus c) == a \otimes b \oplus a \otimes c}$
True
$\boldsymbol{a \oplus (b \otimes c) == (a \oplus b) \otimes (a \oplus c)}$
True

Idempotent law:
$\boldsymbol{a \otimes a == a}$
True
$\boldsymbol{a \oplus a == a}$
True

Law of absorption:
$\boldsymbol{a \otimes (a \oplus b) == a}$
True
$\boldsymbol{a \oplus (a \otimes b) == a}$
True

Complementation:
$\boldsymbol{a \otimes (1 - a) == 0}$
True
$\boldsymbol{a \oplus (1 - a) == 1}$
True

De Morgan's theorem:
$\boldsymbol{(1 - (a \otimes b)) == (1 - a) \oplus (1 - b)}$
True
$\boldsymbol{(1 - (a \oplus b)) == (1 - a) \otimes (1 - b)}$
True

The ⊖ and ⊗ operators have been defined so that they operate on any number of arguments. Consider the following operations involving **`CirclePlus:`**

```
CirclePlus[a, b, c, d]
a + b − ab + c − ac − bc + abc + d − ad − bd + abd − cd + acd + bcd − abcd
```

This result is equivalent to the expression **(a ⊕ b ⊕ c ⊕ d):**

```
CirclePlus[a, b, c, d] == (a ⊕ b ⊕ c ⊕ d)
True
```

The expression **CirclePlus[a, b, c, d]** is an example corresponding to an at least 1-out-of-4:G system:

```
CirclePlus[a, b, c, d] == RaL[1, {a, b, c, d}]
True
```

Similarly, for the following operation involving **CircleTimes,**

```
CircleTimes[a, b, c, d]
abcd
```

This result is also equivalent to **(a ⊗ b ⊗ c ⊗ d):**

```
CircleTimes[a, b, c, d] == (a ⊗ b ⊗ c ⊗ d)
True
```

The expression **CircleTimes[a, b, c, d]** is an example corresponding to an at least or exactly 4-out-of-4:G system:

```
CircleTimes[a, b, c, d] == RaL[4, {a, b, c, d}] == ReX[4, {a, b, c, d}]
True
```

Of course, the use of the BSV technique illustrated here and in Section 3.3 does not avoid the consequences of the NP-complete nature of the reliability problem.[2] As the number of system components increases, the size of the resulting fully expanded reliability polynomial can, in general, increase exponentially in length. This dramatic increase in expanded polynomial length ultimately limits the maximum size of the system that can be evaluated using the BSV technique to about two dozen components. Nevertheless, *Mathematica* can work effectively with rather large polynomials, and within these limits, the technique can be of significant use. In many cases, large systems can be modeled by breaking the overall system into a series of

---

[2] It would be extremely useful if there existed an algorithm that, when applied to a factored polynomial, produced a modified expression (still in factored form) that would be algebraically equivalent to the application of **BernoulliRule** to the fully expanded polynomial. If such an algorithm existed, reliability problems involving large numbers of components could be easily handled symbolically, but insofar as the general problem is known to be NP-complete, the existence of such an algorithm is unlikely.

independent conditional probability models that can then be combined to represent the reliability of a much larger system. In subsequent chapters, BSV techniques are used in the development of these conditional probability models.

Chapter 4
Imperfect Fault Coverage

Hey man—what's the problem?

Abstract Redundant systems must include some means by which they can detect, isolate and reconfigure components in the event of failures; this process is often referred to as *redundancy management*. In real-world systems, this redundancy management task can seldom be done with perfect certainty, and consequently, such systems are said to be subject to *imperfect fault coverage*. Imperfect fault coverage can have a significant impact on system reliability and must be included in system reliability modeling to obtain correct results. The correct modeling of imperfect fault coverage depends on the specifics of the system's architecture and redundancy management approach. This chapter provides various techniques for handling the effects of imperfect fault coverage in the assessment of redundant system reliability.

4.1 Background

Computer-controlled systems that are used in life-critical applications must often employ redundancy to meet the required levels of reliability. An increasing interest in the development of highly reliable systems has been a significant reason for the extensive treatment in the literature of reliability models for k-out-of-n:G systems. Most of this literature, however, only treats the perfect fault coverage (PFC) case, meaning that the system is perfectly capable of detecting, isolating and accommodating failures among the redundant elements. All redundant systems must have a means of accomplishing the tasks of fault detection, isolation and accommodation; this system function is called *redundancy management* (RM). In actual systems, the RM task can seldom, if ever, be performed with complete certainty. Consequently, these systems are subject to imperfect fault coverage (IFC). Even for highly reliable systems with coverage levels approaching unity, the lack of perfect fault coverage still has a significant adverse effect on the probability of system failure. As a result, appropriate modeling of the effects of coverage is critical to the design of these systems—particularly those with operation that is life critical or has a substantial financial consequence.

This chapter presents techniques that employ combinatorial, recursive and table-based functions to calculate the reliability of k-out-of-n:G systems subject to imperfect fault coverage. There are two conceptual approaches to the design and, consequently, the modeling of systems subject to IFC: fault coverage can be modeled as a function of the number of faults that the system has experienced, which is fault level coverage (FLC), or it can be assumed that a particular coverage value can be associated with each redundant element in the system, which is element level coverage (ELC). The IFC taxonomy presented hereafter is based on that used in [1].

FLC is appropriate for modeling systems in which the selection among redundant elements varies between initial and subsequent failures. Systems with RM logic that is based on the comparison of the outputs, or voting of the redundant elements, are candidates for FLC. A system with three or more redundant elements can be designed to ensure extremely high levels of coverage as long as a mid-value-select voting strategy can be applied; if the outputs of the last two remaining elements do not agree, however, then the selection between them cannot be made with the same high coverage level. One-on-one level coverage (OLC) is a special case of FLC that treats faults prior to the one-on-one fault as having perfect coverage, or a coverage value of unity. For systems with $n \geq 3$, it may be adequate to consider coverage for only the one-on-one fault condition.

ELC is probably most appropriate if the selection among the redundant elements is made on the basis of a self-diagnostic capability within the individual elements. That is, each redundant element may, in addition to its primary output, also have an output that indicates the operational status of the element. Such systems typically contain a built-in test (BIT) capability. If ELC is used to model the reliability of a system with redundant elements that use BIT, then the coverage value for a particular element can be modeled as the product of the reliability of the BIT system and the probability that the BIT will correctly identify the failure of its associated element. To the extent that the reliability literature has addressed the topic of IFC at all prior to the publication of [1], it has usually addressed only ELC systems.[1]

The results presented in this book do not distinguish between permanent, intermittent and transient faults, because the system will fail if the RM approach fails to protect the system's operation from incorrect redundant element outputs, whatever the reason. Nevertheless, RM techniques that provide robust protection from non-permanent faults do exist, and they are also able to restore the temporarily faulted element to the active redundant set of the system. The RM approach described during the discussion of FLC systems is an example.

The combinatorial techniques presented in Section 4.4 are particularly well suited for producing symbolic results, which are useful for gaining insight into the functional relationships that determine system reliability. The recursive functions presented in Section 4.6 and the table-based algorithms presented in Section 4.7 yield

---

[1] This does not mean, however, that analysis of systems using FLC or OLC techniques was not done prior to this time; for example, conditional probability models using OLC were employed in the probability of loss-of-control analysis of the digital flight control system for the B-2 bomber in the early 1980s.

results identical to those of the combinatorial models. Of these three modeling approaches, the table-based functions are the most computationally efficient.

Although Chapters 2 and 3 treat the reliability of systems that have a perfect ability to manage their redundant resources (also known as PFC systems), the present chapter covers the reliability modeling of redundant systems subject to IFC.

The recursive functions and table-based algorithms presented in this chapter are implemented using the $\otimes$ and $\oplus$ operators, which are discussed in Section 3.5. Consequently, these functions and algorithms produce correct results for k-out-of-n:G systems even if the redundant elements are general reliability polynomials that are not necessarily disjoint.

4.2 Imperfect Fault Coverage Models

This section treats various IFC models, or k-out-of-n:G systems that use ELC, FLC or OLC models. Recall that a k-out-of-n:G system is a system that has n redundant (but not necessarily identical) components and that is operational if at least k out of the n components remain functional.

4.2.1 ELC Systems

An ELC system is a k-out-of-n:G system for which each component p_i $(i = 1, \ldots, n)$ has a coverage value c_i (that is, a probability of individual component fault coverage). The system is failed if any p_i is failed uncovered, or if more than $n - k$ of the components p_i are failed.

ELC is appropriate if the RM selection among the redundant elements is made based on a self-diagnostic capability in the individual elements. That is, the redundant elements have, in addition to their primary output, a status output that indicates the operational status of the element. If ELC is used to model the reliability of a system with redundant elements that use BIT, then the coverage value for a particular element can be modeled as the product of the reliability of the BIT system and the probability that the BIT will correctly identify the failure of its associated element.

The level of ELC that can be achieved in actual systems depends on the time required by the RM process. For systems that can run the RM process with a periodicity measured on the order of several seconds or minutes (or longer), BIT reliability and (as a result) ELC coverage level may even be greater than 99%, depending on the nature of the element BIT. Nevertheless, if a system requires that the RM task be performed multiple times per second, such as is required for aircraft digital flight control systems, then component BIT generally cannot be performed with a confidence greater than 90%–99%. These differences in BIT confidence are due to the impossibility of performing exhaustive BIT in very short times. That is, given enough time, the BIT can test a greater fraction of the possible failure modes of the system.

Systems that use ELC-based RM architectures are generally not capable of meeting extremely stringent reliability requirements, such as those required for aircraft flight control systems, because the level of achievable coverage (generally 99% or less) severely limits the level of reliability achievable through the use of redundancy.

4.2.2 FLC Systems

An FLC system is a k-out-of-n:G system that organizes a vote among components $p_1, p_2, \ldots, p_n$. The first failure has a probability c_1 of being covered, the second failure has a probability c_2 of being covered and so on, up to c_{n-k}. The $(n-k+1)$th failure will cause the system to fail regardless of whether the failure is covered. Again, the system is failed if any p_i fails uncovered or if fewer than k of the components p_i are operational.

FLC is appropriate for modeling systems in which the fault detection and reconfiguration RM tasks vary between initial and subsequent failures. If the system includes fault detection and reconfiguration logic that is based on comparison of the outputs (voting), then the redundant elements are candidates for FLC modeling. A system with three or more redundant elements can be designed to assure extremely high levels of coverage as long as a mid-value-select (MVS) voting strategy can be applied. Nevertheless, selection between the last two remaining elements, which have outputs that disagree by an amount in excess of some predetermined fault detection threshold, cannot be made with the same high coverage level. Note that this lower coverage level is *not* due to an inability to detect the fault; rather, it is due to an inability to determine which of the two elements with disparate outputs is the failed element. RM for this one-on-one case is often accomplished primarily using BIT, as is frequently done for all ELC system failures, perhaps augmented with heuristics that are based on the nature of the element. Thus, initial faults (those that occur when the redundant set still has three or more elements of a given type) are subject to an FLC near unity.

On the other hand, after the system has failed down to a "one-on-one" configuration (where, for the $k = 1$ case, only two components remain operational), the coverage for the last sustainable fault is no better than the coverage that is typically associated with ELC systems. For FLC systems, coverage for the initial faults is close to unity, and only the one-on-one fault has a coverage level that is typical of an ELC system. As a result, FLC systems can be designed to achieve much lower probabilities of failure. For this reason, most manned military digital aircraft flight control systems, which are usually designed to have a probability of failure on the order of 5×10^{-7} per mission, are designed as FLC systems.

4.2.3 OLC Systems

Since the coverage levels for the initial faults in FLC systems can be close to unity, these FLC systems can frequently be modeled with sufficient accuracy by assuming that the initial coverage values are, in fact, equal to unity. This simplification can significantly reduce the complexity of the system reliability calculations and the length of the reliability polynomial. A model that uses this approximation for one-on-one level coverage is called *OLC*.

4.3 IFC Sum-over-States Models

The following discussion includes the use of the sum-over-states (truth-table) technique to determine system reliability for 1-out-of-3:G ELC and FLC systems. In these examples, the reliability of the redundant components are represented by the vector $\mathbf{p} = \{p_1, \ldots, p_n\}$. For the ELC case, the elements of the coverage vector $\mathbf{c} = \{c_1, \ldots, c_n\}$ represent the probabilities that the corresponding components will fail covered; that is, c_i is the coverage for the ith component. The FLC model has a coverage vector $\mathbf{c} = \{c_1, \ldots, c_{n-1}\}$, where c_i is the probability that the ith failure among the n redundant components is covered. Although both models use the same c_i nomenclature for coverage values, bear in mind that the values represent different probabilities, depending on whether the model is for an ELC or FLC system.

For an ELC system, the probability that a given system state is operational depends on which of the redundant components have failed: if Component 1 fails, the failure will be covered with probability c_1; if Component 2 fails, the failure will be covered with probability c_2; if both Components 1 and 2 have failed, the coverage will be c_1c_2. Other cases follow this pattern. Table 4.1 is a truth table for all possible failure combinations of a 1-out-of-3:G ELC system. The reliability of a 1-out-of-3:G ELC system is the sum of the entries in the P(OP) column.

For an FLC system, the probability that a given system state is operational is a function of the number of covered failures that the system has experienced; if the system has experienced one failure, then there is a probability c_1 that the failure was covered. For two failures, the probability that both were covered is c_1c_2. Table 4.2 shows all of the possible failure combinations for a 1-out-of-3:G FLC system. Again, the 1-out-of-3:G FLC system reliability is the sum of the P(OP) entries.

Table 4.1 Possible states of a 1-out-of-3:G ELC system

State	p_1	p_2	p_3	P(state)	P(OP)
1	p_1	p_2	p_3	$p_1p_2p_3$	$p_1p_2p_3$
2	p_1	p_2	q_3	$p_1p_2q_3$	$c_3p_1p_2q_3$
3	p_1	q_2	p_3	$p_1q_2p_3$	$c_2p_1q_2p_3$
4	q_1	p_2	p_3	$q_1p_2p_3$	$c_1q_1p_2p_3$
5	p_1	q_2	q_3	$p_1q_2q_3$	$c_2c_3p_1q_2q_3$
6	q_1	p_2	q_3	$q_1p_2q_3$	$c_1c_3q_1p_2q_3$
7	q_1	q_2	p_3	$q_1q_2p_3$	$c_1c_2q_1q_2p_3$
8	q_1	q_2	q_3	$q_1q_2q_3$	0

Table 4.2 Possible states of a 1-out-of-3:G FLC system

State	p_1	p_2	p_3	P(state)	P(OP)
1	p_1	p_2	p_3	$p_1p_2p_3$	$p_1p_2p_3$
2	p_1	p_2	q_3	$p_1p_2q_3$	$c_1p_1p_2q_3$
3	p_1	q_2	p_3	$p_1q_2p_3$	$c_1p_1q_2p_3$
4	q_1	p_2	p_3	$q_1p_2p_3$	$c_1q_1p_2p_3$
5	p_1	q_2	q_3	$p_1q_2q_3$	$c_1c_2p_1q_2q_3$
6	q_1	p_2	q_3	$q_1p_2q_3$	$c_1c_2q_1p_2q_3$
7	q_1	q_2	p_3	$q_1q_2p_3$	$c_1c_2q_1q_2p_3$
8	q_1	q_2	q_3	$q_1q_2q_3$	0

The sum of the P(OP) entries from Table 4.1 is the reliability of a 1-out-of-3:G ELC system, as expressed in Equation (4.1).

$$\begin{aligned}
\mathrm{R_{ELC}} &= p_1p_2p_3 + c_1p_2p_3q_1 + c_2p_1p_3q_2 + c_1c_2p_3q_1q_2 \\
&\quad + c_3p_1p_2q_3 + c_1c_3p_2q_1q_3 + c_2c_3p_1q_2q_3 \\
&= p_1p_2p_3 + c_1p_2p_3(1-p_1) + c_2p_1p_3(1-p_2) \\
&\quad + c_1c_2p_3(1-p_1)(1-p_2) + c_3p_1p_2(1-p_3) \\
&\quad + c_1c_3p_2(1-p_1)(1-p_3) + c_2c_3p_1(1-p_2)(1-p_3) \qquad (4.1)\\
&= c_2c_3p_1 + c_1c_3p_2 + c_3p_1p_2 - c_1c_3p_1p_2 - c_2c_3p_1p_2 \\
&\quad + c_1c_2p_3 + c_2p_1p_3 - c_1c_2p_1p_3 - c_2c_3p_1p_3 + c_1p_2p_3 \\
&\quad - c_1c_2p_2p_3 - c_1c_3p_2p_3 + p_1p_2p_3 - c_1p_1p_2p_3 - c_2p_1p_2p_3 \\
&\quad + c_1c_2p_1p_2p_3 - c_3p_1p_2p_3 + c_1c_3p_1p_2p_3 + c_2c_3p_1p_2p_3
\end{aligned}$$

The corresponding expression from Table 4.2 yields the 1-out-of-3:G FLC system reliability, as expressed in Equation (4.2).

$$
\begin{aligned}
R_{\text{FLC}} &= p_1p_2p_3 + c_1p_2p_3q_1 + c_1p_1p_3q_2 + c_1c_2p_3q_1q_2 \\
&\quad + c_1p_1p_2q_3 + c_1c_2p_2q_1q_3 + c_1c_2p_1q_2q_3 \\
&= p_1p_2p_3 + c_1p_2p_3(1-p_1) + c_1p_1p_3(1-p_2) \\
&\quad + c_1c_2p_3(1-p_1)(1-p_2) + c_1p_1p_2(1-p_3) \\
&\quad + c_1c_2p_2(1-p_1)(1-p_3) + c_1c_2p_1(1-p_2)(1-p_3) \\
&= c_1c_2p_1 + c_1c_2p_2 + c_1p_1p_2 - 2c_1c_2p_1p_2 \\
&\quad + c_1c_2p_3 + c_1p_1p_3 - 2c_1c_2p_1p_3 + c_1p_2p_3 \\
&\quad - 2c_1c_2p_2p_3 + p_1p_2p_3 - 3c_1p_1p_2p_3 + 3c_1c_2p_1p_2p_3
\end{aligned}
\tag{4.2}
$$

Sections 4.4, 4.6 and 4.7 present combinatorial, recursive and table-based algorithms for computing k-out-of-n:G reliability for ELC, FLC and OLC systems. These functions generate reliability expressions that are identical to the results computed above using sum over states.

4.4 IFC Combinatorial Functions

The reliability of ELC, FLC and OLC k-out-of-n:G systems with non i.i.d. element reliabilities and with redundant component reliability vector $\mathbf{p}$ can be computed using the functions presented in this section along with functions for IFC i.i.d. k-out-of-n:G systems. These models are the IFC counterparts of the PFC k-out-of-n:G function in Equation (2.9). The combinatorial IFC functions presented in this section are from [1]. The following sets are used in these combinatorial IFC functions:

- $\mathbf{pT}(i, \mathbf{p})$ Set of products of the k-subsets of $\mathbf{p}$ with exactly i elements
- $\mathbf{qT}(i, \mathbf{q})$ Set of products of the k-subsets of $\mathbf{q}$ with exactly $n-i$ elements
- $\mathbf{cT}(i, \mathbf{c})$ Set of products of the k-subsets of $\mathbf{c}$ with exactly $n-i$ elements, where $\mathbf{c}$ = ELC coverage vector ($|\mathbf{c}| = n$)

Note that the k-subsets must be generated in lexicographic order. The $\mathbf{pT}$ and $\mathbf{qT}$ sets are the same as those described in Section 2.8.1. The set $\mathbf{cT}$ is generated in a similar manner.

4.4.1 ELC Functions

The functions for generating ELC at least and exactly k-out-of-n:G reliabilities involve straightforward modifications of Equations (2.9) and (2.11). ELC at least k-out-of-n:G reliability can be computed using Equation (4.3), and ELC exactly k-out-of-n:G reliability can be computed using Equation (4.4).

$$R_{ELC}(k,n,\mathbf{p},\mathbf{c}) = \sum_{i=k}^{n}\sum_{j=1}^{\binom{n}{i}} \mathbf{cT}(i,\mathbf{c})_j \mathbf{qT}(i,\mathbf{p})_j \mathbf{pT}(i,\mathbf{p})_{\binom{n}{i}-j+1} \tag{4.3}$$

$$Re_{ELC}(k,n,\mathbf{p},\mathbf{c}) = \sum_{j=1}^{\binom{n}{k}} \mathbf{cT}(k,\mathbf{c})_j \mathbf{qT}(k,\mathbf{p})_j \mathbf{pT}(k,\mathbf{p})_{\binom{n}{k}-j+1} \tag{4.4}$$

For example, given $\mathbf{p} = \{p_1, p_2, p_3\}$ and $\mathbf{c} = \{c_1, c_2, c_3\}$ such that $n = |\mathbf{p}| = 3$, the required **pT, qT** and **cT** sets are

$$\begin{aligned}
\mathbf{pT}(1,\mathbf{p}) &= \{p_1, p_2, p_3\}\\
\mathbf{pT}(2,\mathbf{p}) &= \{p_1p_2, p_1p_3, p_2p_3\}\\
\mathbf{pT}(3,\mathbf{p}) &= \{p_1p_2p_3\}
\end{aligned}$$

$$\begin{aligned}
\mathbf{qT}(1,\mathbf{p}) &= \{(1-p_1)(1-p_2), (1-p_1)(1-p_3), (1-p_2)(1-p_3)\}\\
\mathbf{qT}(2,\mathbf{p}) &= \{(1-p_1), (1-p_2), (1-p_3)\}\\
\mathbf{qT}(3,\mathbf{p}) &= 1
\end{aligned}$$

$$\begin{aligned}
\mathbf{cT}(1,\mathbf{c}) &= \{c_1c_2, c_1c_3, c_2c_3\}\\
\mathbf{cT}(2,\mathbf{c}) &= \{c_1, c_2, c_3\}\\
\mathbf{cT}(3,\mathbf{c}) &= \{1\} \ .
\end{aligned}$$

For ELC at least 1-out-of-3:G reliability,

$$\begin{aligned}
R_{ELC}(1,3,\mathbf{p},\mathbf{c}) = {}& \\
& (c_1(-1+p_1) - p_1)\, p_2\, (c_3(-1+p_3) - p_3)\\
& + c_2(-1+p_2)(c_3p_1(-1+p_3) + (c_1(-1+p_1) - p_1)\, p_3)\\
={}& c_2c_3p_1 + c_1c_3p_2 + c_3p_1p_2 - c_1c_3p_1p_2 - c_2c_3p_1p_2\\
& + c_1c_2p_3 + c_2p_1p_3 - c_1c_2p_1p_3 - c_2c_3p_1p_3 + c_1p_2p_3\\
& - c_1c_2p_2p_3 - c_1c_3p_2p_3 + p_1p_2p_3 - c_1p_1p_2p_3 - c_2p_1p_2p_3\\
& + c_1c_2p_1p_2p_3 - c_3p_1p_2p_3 + c_1c_3p_1p_2p_3 + c_2c_3p_1p_2p_3 \ .
\end{aligned} \tag{4.5}$$

For ELC exactly 1-out-of-3:G reliability,

$$\begin{aligned}
Re_{ELC}(1,3,\mathbf{p},\mathbf{c}) = {}& \\
& c_1c_3(-1+p_1)\, p_2\, (-1+p_3)\\
& + c_2(-1+p_2)(c_3p_1(-1+p_3) + c_1(-1+p_1)\, p_3)\\
={}& c_2c_3p_1 + c_1c_3p_2 - c_1c_3p_1p_2 - c_2c_3p_1p_2\\
& + c_1c_2p_3 - c_1c_2p_1p_3 - c_2c_3p_1p_3 - c_1c_2p_2p_3\\
& - c_1c_3p_2p_3 + c_1c_2p_1p_2p_3 + c_1c_3p_1p_2p_3 + c_2c_3p_1p_2p_3 \ .
\end{aligned} \tag{4.6}$$

By applying the **BernoulliRule** substitution to the fully expanded calculations, the R_{ELC} and Re_{ELC} functions yield correct reliability results for k-out-of-n:G systems with redundant inputs that are general reliability polynomials. That is, these results are correct for complex systems. *Mathematica* functions implementing Equations (4.3) and (4.4) are given in Appendix A.2.

4.4.2 FLC Functions

The functions for generating FLC at least and exactly k-out-of-n:G reliabilities involve straightforward modifications of Equations (2.9) and (2.11). The FLC coverage vector has a length $n - 1$.

$$\mathrm{cP}(k, n, \mathbf{c}) = \prod_{i=1}^{n-k} c_i \tag{4.7}$$

$$R_{FLC}(k, n, \mathbf{p}, \mathbf{c}) = \sum_{i=k}^{n} \mathrm{cP}(i, n, \mathbf{c}) \sum_{j=1}^{\binom{n}{i}} \mathbf{qT}(i, \mathbf{p})_j \mathbf{pT}(i, \mathbf{p})_{\binom{n}{i}-j+1} \tag{4.8}$$

$$Re_{FLC}(k, n, \mathbf{p}, \mathbf{c}) = \mathrm{cP}(k, n, \mathbf{c}) \sum_{j=1}^{\binom{n}{k}} \mathbf{qT}(k, \mathbf{p})_j \mathbf{pT}(k, \mathbf{p})_{\binom{n}{k}-j+1} \tag{4.9}$$

For example, given $\mathbf{p} = \{p_1, p_2, p_3\}$ and $\mathbf{c} = \{c_1, c_2\}$ such that $n = |\mathbf{p}| = 3$, the required $\mathbf{pT}$ and $\mathbf{qT}$ sets are

$$\begin{aligned}
\mathbf{pT}(1, \mathbf{p}) &= \{p_1, p_2, p_3\} \\
\mathbf{pT}(2, \mathbf{p}) &= \{p_1 p_2, p_1 p_3, p_2 p_3\} \\
\mathbf{pT}(3, \mathbf{p}) &= \{p_1 p_2 p_3\}
\end{aligned}$$

$$\begin{aligned}
\mathbf{qT}(1, \mathbf{p}) &= \{(1 - p_1)(1 - p_2), (1 - p_1)(1 - p_3), (1 - p_2)(1 - p_3)\} \\
\mathbf{qT}(2, \mathbf{p}) &= \{(1 - p_1), (1 - p_2), (1 - p_3)\} \\
\mathbf{qT}(3, \mathbf{p}) &= \{1\} \ .
\end{aligned}$$

For FLC at least 1-out-of-3:G reliability,

$$\begin{aligned}
R_{FLC}(1, 3, \mathbf{p}, \mathbf{c}) &= p_1 p_2 p_3 + c_1 (p_2 p_3 + p_1 (p_2 + p_3 - 3 p_2 p_3) + \\
&\quad c_2 (p_2 + p_3 - 2 p_2 p_3 + p_1 (1 - 2 p_3 + p_2 (-2 + 3 p_3)))) \\
&= c_1 c_2 p_1 + c_1 c_2 p_2 + c_1 p_1 p_2 - 2 c_1 c_2 p_1 p_2 \\
&\quad + c_1 c_2 p_3 + c_1 p_1 p_3 - 2 c_1 c_2 p_1 p_3 + c_1 p_2 p_3 \\
&\quad - 2 c_1 c_2 p_2 p_3 + p_1 p_2 p_3 - 3 c_1 p_1 p_2 p_3 + 3 c_1 c_2 p_1 p_2 p_3 \ .
\end{aligned} \tag{4.10}$$

For FLC exactly 1-out-of-3:G reliability,

$$\begin{aligned} Re_{FLC}(1,3,\mathbf{p},\mathbf{c}) &= c_1c_2\left(p_2\left(1-2p_3\right)+p_3+p_1\left(1-2p_3+p_2\left(-2+3p_3\right)\right)\right) \\ &= c_1c_2p_1+c_1c_2p_2-2c_1c_2p_1p_2+c_1c_2p_3 \\ &\quad -2c_1c_2p_1p_3-2c_1c_2p_2p_3+3c_1c_2p_1p_2p_3\,. \end{aligned} \tag{4.11}$$

Mathematica functions implementing Equations (4.8) and (4.9) are given in Appendix A.3; again, the **BernoulliRule** substitution is applied to assure that the results are correct for the situation when the elements of $\mathbf{p}$ are general reliability polynomials.

4.4.3 OLC Functions

The functions for generating OLC at least and exactly k-out-of-n:G reliabilities also involve straightforward modifications of Equations (2.9) and (2.11). An OLC system has a single coverage value, c, corresponding to the "one-on-one," or $(n-1)$th, failure.

$$\mathrm{C}(i,c)=\begin{cases} c \text{ if } i=1 \\ 1 \text{ otherwise} \end{cases} \tag{4.12}$$

$$R_{OLC}(k,n,\mathbf{p},\mathbf{c})=\sum_{i=k}^{n}\mathrm{C}(i,c)\sum_{j=1}^{\binom{n}{i}}\mathbf{qT}(i,\mathbf{p})_j\mathbf{pT}(i,\mathbf{p})_{\binom{n}{i}-j+1} \tag{4.13}$$

$$Re_{OLC}(k,n,\mathbf{p},\mathbf{c})=\mathrm{C}(k,c)\sum_{j=1}^{\binom{n}{k}}\mathbf{qT}(k,\mathbf{p})_j\mathbf{pT}(k,\mathbf{p})_{\binom{n}{k}-j+1} \tag{4.14}$$

For example, given $\mathbf{p}=\{p_1,p_2,p_3\}$ and $\mathbf{c}=\{c_1,c_2\}$ such that $n=|\mathbf{p}|=3$, the required $\mathbf{pT}$ and $\mathbf{qT}$ sets are

$$\begin{aligned} \mathbf{pT}(1,\mathbf{p}) &= \{p_1,p_2,p_3\} \\ \mathbf{pT}(2,\mathbf{p}) &= \{p_1p_2,p_1p_3,p_2p_3\} \\ \mathbf{pT}(3,\mathbf{p}) &= \{p_1p_2p_3\} \end{aligned}$$

$$\begin{aligned} \mathbf{qT}(1,\mathbf{p}) &= \{(1-p_1)(1-p_2),(1-p_1)(1-p_3),(1-p_2)(1-p_3)\} \\ \mathbf{qT}(2,\mathbf{p}) &= \{(1-p_1),(1-p_2),(1-p_3)\} \\ \mathbf{qT}(3,\mathbf{p}) &= \{1\}\,. \end{aligned}$$

For OLC at least 1-out-of-3:G reliability,

$$\begin{aligned}R_{OLC}(1,3,\mathbf{p},c) = \\ &p_1p_2(1-p_3)+p_1(1-p_2)p_3+(1-p_1)p_2p_3+p_1p_2p_3 \\ &+c(p_1(1-p_2)(1-p_3)+(1-p_1)p_2(1-p_3) \\ &\quad+(1-p_1)(1-p_2)p_3) \\ &= cp_1+cp_2+p_1p_2-2cp_1p_2+cp_3+p_1p_3 \\ &\quad-2cp_1p_3+p_2p_3-2cp_2p_3-2p_1p_2p_3+3cp_1p_2p_3\,.\end{aligned} \tag{4.15}$$

For OLC exactly 1-out-of-3:G reliability,

$$\begin{aligned}Re_{OLC}(1,3,\mathbf{p},c) &= c(p_1(1-p_2)(1-p_3)+(1-p_1)p_2(1-p_3) \\ &\quad+(1-p_1)(1-p_2)p_3) \\ &= cp_1+cp_2-2cp_1p_2+cp_3-2cp_1p_3-2cp_2p_3 \\ &\quad+3cp_1p_2p_3\,.\end{aligned} \tag{4.16}$$

Mathematica functions implementing Equations (4.13) and (4.14) are given in Appendix A.4.

4.5 Combinatorial Functions for i.i.d. Systems

The i.i.d. PFC functions discussed in Chapter 2—Equations (2.8) and (2.12)—can be modified to provide results for i.i.d. IFC system reliability as well. Since the redundant components of an i.i.d. ELC system are identical, the coverage values for each component will also be identical and can be represented by a scalar value, c. The coverage values for an i.i.d. FLC system, however, are represented by a vector, $\mathbf{c} = \{c_1, \ldots, c_{n-1}\}$, and the $(n-1)$th failure of an OLC system is represented by a single coverage value, c. The IFC k-out-of-n:G i.i.d. coverage function, $C(i,n)$, which is given in Equation (4.17), has a different representation for each of the models—PFC, ELC, FLC and OLC.

$$C(i,n) = \begin{cases} 1 & PFC \\ c^{n-i} & ELC \\ \prod_{j=1}^{n-i} c_j & FLC \\ \begin{cases} i=1 & c \\ \text{otherwise} & 1 \end{cases} & OLC \end{cases} \tag{4.17}$$

The reliability for an i.i.d. at least k-out-of-n:G system is

$$R_{\text{iid}}(k,n) = \sum_{i=k}^{n} C(i,n)\binom{n}{i}p^i(1-p)^{n-i}\,, \tag{4.18}$$

and for an i.i.d. exactly k-out-of-n:G system,

$$Re_{\text{iid}}(k,n) = C(k,n)\binom{n}{k}p^k(1-p)^{n-k}\ . \tag{4.19}$$

The following is a set of *Mathematica* functions implementing Equations (4.17), (4.18) and (4.19):

```
C[i_,n_,type_]:=Module[{j},
  Switch[type,
    PFC, 1,
    ELC, c^(n-i),
    FLC, ∏_{j=1}^{n-i} c[[j]],
    OLC, If[(i==1), c, 1]
  ]
];

Riid[k_,n_,type_]:=Module[{i},
  ∑_{i=k}^{n} Binomial[n, i]p^i(1 - p)^(n-i)C[i, n, type]
];


REiid[k_,n_,type_]:=Module[{},
Binomial[n, k]p^k(1 - p)^(n-k)C[k, n, type]
];
```

The i.i.d. at least 1-out-4:G system reliability for each of the coverage models is

```
Riid[1,4,PFC]//Expand
```
$4p - 6p^2 + 4p^3 - p^4$

```
Riid[1,4,ELC]//Expand
```
$4c^3p + 6c^2p^2 - 12c^3p^2 + 4cp^3 - 12c^2p^3 + 12c^3p^3 + p^4 - 4cp^4 + 6c^2p^4 - 4c^3p^4$

Define a coverage vector of length $n-1$ for the FLC case:

```
c = {c1, c2, c3};

Riid[1,4,FLC]//Expand
```
$4\text{c1c2c3}p + 6\text{c1c2}p^2 - 12\text{c1c2c3}p^2 + 4\text{c1}p^3 - 12\text{c1c2}p^3 + 12\text{c1c2c3}p^3$
$+ p^4 - 4\text{c1}p^4 + 6\text{c1c2}p^4 - 4\text{c1c2c3}p^4$

Define a scalar coverage value for the OLC case:

```
Clear[c];
```

```
Riid[1,4,OLC]//Expand
```
$4cp + 6p^2 - 12cp^2 - 8p^3 + 12cp^3 + 3p^4 - 4cp^4$

4.6 Recursive k-out-of-n:G Functions

Additional discussion and background on the development of the recursive functions related to those presented in this section can be found in [2]. A discussion of recursive functions for ELC and FLC k-out-of-n:F systems is included in [3], which also explains the relationship between the recursive functions and the corresponding table-based algorithms presented in Section 4.7.

The results obtained using the combinatorial models of Section 4.4 can also be calculated, with modestly improved computational efficiency, using the recursive functions described in this section. These recursive functions also provide the derivational foundation for the table-based algorithms presented in the next section. Recursive functions are presented for both PFC and IFC systems. The recursive functions defined in this section use the $\otimes$ and $\oplus$ operators, which are defined in Section 3.4. As a result, these functions provide correct results even if the vector of redundant inputs, $\mathbf{p}$, contains general reliability polynomials that are not necessarily disjoint. The *Mathematica* functions that implement the recursive functions in this section are given in Appendix B.

4.6.1 PFC Recursive Functions

The reliability of a simple PFC at least k-out-of-n:G system can be computed using the recursive function given in Equation (4.20), subject to the boundary cases given in Equations (4.21) and (4.22). PFC exactly k-out-of-n:G system reliability is given in Equation (4.23), with boundary cases given in Equations (4.24) and (4.25).

Given $\mathbf{p} = \{p_1, \ldots, p_n\}$ and $n = |\mathbf{p}|$,

$$R_{PFC}(k,n) = (1 - p_n) \otimes R_{PFC}(k, n-1) \oplus p_n \otimes R_{PFC}(k-1, n-1)\,. \tag{4.20}$$

This expression is subject to the following boundary cases:

$$R_{PFC}(k,n) = 0 \quad \text{if} \quad n = 0 \quad \text{and} \quad k > n \tag{4.21}$$

$$R_{PFC}(k,n) = 1 \quad \text{if} \quad n = 0 \quad \text{and} \quad k \leq n\,. \tag{4.22}$$

The recursive function returning PFC exactly k-out-of-n:G system reliability has the same recursion formula but uses a different set of boundary cases.

$$Re_{PFC}(k,n) = (1 - p_n) \otimes Re_{PFC}(k, n-1) \oplus p_n \otimes Re_{PFC}(k-1, n-1) \tag{4.23}$$

This expression is subject to the following boundary cases:

$$Re_{PFC}(k,n) = 0 \quad \text{if} \quad n < k \quad \text{or} \quad k < 0 \tag{4.24}$$

$$Re_{PFC}(k,n) = 1 \quad \text{if} \quad n = 0 \quad \text{and} \quad k = n\,. \tag{4.25}$$

Since these recursive functions use the $\oplus$ and $\otimes$ operators defined in Section 3.4, they yield correct results even if the redundant elements of the system are general reliability polynomials that are not necessarily be disjoint. The ELC and FLC recursive functions defined below are implemented in the same fashion.

4.6.2 ELC Recursive Functions

The reliability of a simple ELC at least k-out-of-n:G system can be computed using the recursive function in Equation (4.26), subject to the boundary cases expressed in Equations (4.27) and (4.28). ELC exactly k-out-of-n:G system reliability is given in Equation (4.29), with boundary cases given in Equations (4.30) and (4.31).

Given $\mathbf{p} = \{p_1, \ldots, p_n\}$ and $\mathbf{c} = \{c_1, \ldots, c_n\}$,

$$R_{ELC}(k,n) = (1 - p_n) \otimes c_n \otimes R_{ELC}(k, n-1) \oplus p_n \otimes R_{ELC}(k-1, n-1)\,. \tag{4.26}$$

The above expression is subject to the following boundary cases:

$$R_{ELC}(k,n) = 0 \quad \text{if} \quad n = 0 \quad \text{and} \quad k > n \tag{4.27}$$

$$R_{ELC}(k,n) = 1 \quad \text{if} \quad n = 0 \quad \text{and} \quad k \leq n\,. \tag{4.28}$$

The recursive function returning ELC exactly k-out-of-n:G system reliability has the same recursion formula but uses a different set of boundary cases.

$$\begin{aligned} Re_{ELC}(k,n) = (1 - p_n) \otimes c_n \otimes Re_{ELC}(k, n-1) \\ \oplus\, p_n \otimes Re_{ELC}(k-1, n-1) \end{aligned} \tag{4.29}$$

The above expression is subject to the following boundary cases:

$$Re_{ELC}(k,n) = 0 \quad \text{if} \quad n < k \quad \text{or} \quad k < 0 \tag{4.30}$$

$$Re_{ELC}(k,n) = 1 \quad \text{if} \quad n = 0 \quad \text{and} \quad k = n\,. \tag{4.31}$$

4.6.3 FLC Recursive Functions

Although the coverage vector $\mathbf{c}$ for an FLC system contains only $n - 1$ elements, the recursive functions below require that this vector be augmented with a trailing zero as the nth element. This addition is required to avoid making an out-of-bounds reference to the element c_n (which, actually, does not appear in the result). Additionally, the FLC functions include a third argument, f, that is incremented with each recursive call to indicate the number of components that have failed up to that point. Thus, the system reliability is equal to $R_{FLC}(k, n, 0)$ (at least k-out-of-n:G) and $Re_{FLC}(k, n, 0)$ (exactly k-out-of-n:G), with $f = 0$ indicating that the process is started with zero failures. If the FLC systems are characterized by $\mathbf{p} = \{p_1, \ldots, p_n\}$ and $\mathbf{c} = \{c_1, \ldots, c_{n-1}, 0\}$, then

$$R_{FLC}(k, n, f) = (1 - p_n) \otimes c_{f+1} \otimes R_{FLC}(k, n - 1, f + 1) \\ \oplus p_n \otimes R_{FLC}(k - 1, n - 1, f) \,. \tag{4.32}$$

The above expression is subject to the following boundary cases:

$$R_{FLC}(k, n, f) = 0 \quad \text{if} \quad n = 0 \quad \text{and} \quad k > n \tag{4.33}$$

$$R_{FLC}(k, n, f) = 1 \quad \text{if} \quad n = 0 \quad \text{and} \quad k \leq n \,. \tag{4.34}$$

The recursive function returning FLC exactly k-out-of-n:G system reliability has the same recursion formula but uses a different set of boundary cases.

$$Re_{FLC}(k, n, f) = (1 - p_n) \otimes c_{f+1} \otimes Re_{FLC}(k, n - 1, f + 1) \\ \oplus p_n \otimes Re_{FLC}(k - 1, n - 1, f) \tag{4.35}$$

The above the expression is subject to the following boundary cases:

$$Re_{FLC}(k, n, f) = 0 \quad \text{if} \quad n < k \quad \text{or} \quad k < 0 \tag{4.36}$$

$$Re_{FLC}(k, n, f) = 1 \quad \text{if} \quad n = 0 \quad \text{and} \quad k = n \,. \tag{4.37}$$

4.6.4 OLC Recursive Functions

Although OLC systems are characterized by a single coverage value, c, the most straightforward approach to obtaining OLC results recursively is to use the FLC functions (4.32) and (4.35), along with coverage vector $\mathbf{c} = \{1, 1, \ldots, c, 0\}$. Here, c is the $(n - 1)$th element of **c,** which is a vector of length n. Alternatively, the *Mathematica* implementation given in Appendix B provides a recursive implementation for the OLC functions where the OLC coverage value is defined as a scalar value, c.

4.7 PFC and IFC Table-Based Algorithms

Using the recursive functions from Section 4.6 as the foundation and following a scheme of "memorization" of the intermediate results, table-based algorithms for the calculation of both PFC and IFC k-out-of-n:G reliabilities can be derived. These algorithms have computational efficiencies superior to both the combinatorial and recursive formulations, especially for systems with large n. With a computational efficiency of $O(n \cdot (n - k))$, these codes are among the most efficient algorithms for the calculation of k-out-of-n:G system reliability. This section provides table-based algorithms for each of the classes of k-out-of-n:G systems that were covered previously, including both PFC and IFC.

The table-based IFC k-out-of-n:G system reliability algorithms presented in this section are reported in [2]. Additional discussion and background on the development of table-based functions closely related to those presented in this section (involving k-out-of-n:F algorithms), can be found in [3]. A *Mathematica* implementation of the k-out-of-n:G algorithms presented in this section is given in Appendix C.

4.7.1 PFC Table-Based Algorithms

The reliability of PFC at least and exactly k-out-of-n:G systems can be computed using the table-based functions R_{PFC} (see Figure 4.1) and Re_{PFC} (see Figure 4.2). The functions R_{PFC} and Re_{PFC} have as input arguments k, the number of components that must be operational, and $\mathbf{p} = \{p_1, \ldots, p_n\}$, the vector of redundant component reliabilities.

```
R_PFC[k, p]
    n ← length[p]
    P ← array of length[n − k + 1]
    P[1] ← 1
    r ← 0
    for i = 2 upto n − k + 1 do P[i] ← 0 done

    for i = 1 upto n do
      for j = n − k downto 1 do
        P[j + 1] ← p_i ⊗ P[j + 1] ⊕ (1 − p_i) ⊗ P[j]
        if i == n then r ← r ⊕ P[j + 1]
      done
      P[1] ← p_i ⊗ P[1]
    done

    return r ⊕ P[1]
```

Fig. 4.1 Table-based PFC at least k-out-of-n:G algorithm

```
Re_PFC[k, p]
    n ← length[p]
    P ← array of length[n − k + 1]
    P[1] ← 1
    for i = 2 upto n − k + 1 do P[i] ← 0 done

    for i = 1 upto n do
      for j = n − k downto 1 do
        P[j + 1] ← p_i ⊗ P[j + 1] ⊕ (1 − p_i) ⊗ P[j]
        if i == n then return P[j + 1]
      done
      P[1] ← p_i ⊗ P[1]
    done

    return P[1]
```

Fig. 4.2 Table-based PFC exactly k-out-of-n:G algorithm

4.7.2 ELC Table-Based Algorithms

The reliabilities of ELC at least and exactly k-out-of-n:G systems can be computed using the table-based functions R_{ELC} (see Figure 4.3) and Re_{ELC} (see Figure 4.4). In addition to k and **p**, the functions R_{ELC} and Re_{ELC} have an input argument $\mathbf{c} = \{c_1, \ldots, c_n\}$, which includes the coverage values associated with each of the components p_i.

```
R_ELC[k, p, c]
    n ← length[p]
    P ← array of length[n − k + 1]
    P[1] ← 1
    r ← 0
    for i = 2 upto n − k + 1 do P[i] ← 0 done

    for i = 1 upto n do
      for j = n − k downto 1 do
        P[j + 1] ← p_i ⊗ P[j + 1] ⊕ (1 − p_i) ⊗ c_i ⊗ P[j]
        if i == n then r ← r ⊕ P[j + 1]
      done
      P[1] ← p_i ⊗ P[1]
    done

    return r ⊕ P[1]
```

Fig. 4.3 Table-based ELC at least k-out-of-n:G algorithm

```
Re_ELC[k, p, c]
    n ← length[p]
    P ← array of length[n − k + 1]
    P[1] ← 1
    for i = 2 upto n − k + 1 do P[i] ← 0 done

    for i = 1 upto n do
      for j = n − k downto 1 do
        P[j + 1] ← p_i ⊗ P[j + 1] ⊕ (1 − p_i) ⊗ c_i ⊗ P[j]
        if i == n then return P[j + 1]
      done
      P[1] ← p_i ⊗ P[1]
    done

    return P[1]
```

Fig. 4.4 Table-based ELC exactly k-out-of-n:G algorithm

4.7.3 FLC Table-Based Algorithms

The reliabilities of FLC at least and exactly k-out-of-n:G systems can be computed using the table-based functions R_{FLC} (see Figure 4.5) and Re_{FLC} (see Figure 4.6). The R_{FLC} and Re_{FLC} arguments include k, $\mathbf{p}$ and the vector $\mathbf{c} = \{c_1, \ldots, c_{n-1}\}$, which contains the coverage values for the 1st through $(n - 1)$th failures among the n redundant components.

```
R_FLC[k, p, c]
    n ← length[p]
    P ← array of length[n − k + 1]
    P[1] ← 1
    r ← 0
    for i = 2 upto n − k + 1 do P[i] ← 0 done

    for i = 1 upto n do
      for j = n − k downto 1 do
        P[j + 1] ← p_i ⊗ P[j + 1] ⊕ (1 − p_i) ⊗ c_j ⊗ P[j]
        if i == n then r ← r ⊕ P[j + 1]
      done
      P[1] ← p_i ⊗ P[1]
    done

    return r ⊕ P[1]
```

Fig. 4.5 Table-based FLC at least k-out-of-n:G algorithm

```
Re_FLC[k, p, c]
    n ← length[p]
    P ← array of length[n − k + 1]
    P[1] ← 1
    for i = 2 upto n − k + 1 do P[i] ← 0 done

    for i = 1 upto n do
      for j = n − k downto 1 do
        P[j + 1] ← p_i ⊗ P[j + 1] ⊕ (1 − p_i) ⊗ c_j ⊗ P[j]
        if i == n then return P[j + 1]
      done
      P[1] ← p_i ⊗ P[1]
    done

    return P[1]
```

Fig. 4.6 Table-based FLC exactly k-out-of-n:G algorithm

Note that the table-based FLC and ELC algorithms are nearly identical; they differ only in the definition of the coverage vector, **c**, and in the subscript in the accumulation term, which changes from i for the ELC case to j for the FLC case.

4.7.4 OLC Table-Based Algorithms

The reliability of OLC at least and exactly k-out-of-n:G systems can be computed using the table-based functions R_{OLC} (see Figure 4.7) and Re_{OLC} (see Figure 4.8). The R_{OLC} and Re_{OLC} functions include a scalar coverage value, c, for the $(n-1)$th failure.

4.8 Estimation of FLC Coverage

Systems in which coverage varies during a sequence of faults are appropriately modeled as FLC systems. The technique used to estimate the $n-1$ coverage values required for an FLC system must be based on the characteristics of the system's RM architecture. An RM scheme representative of some military aircraft digital flight control systems is summarized in Figure 4.9.

An estimate of the values of an FLC coverage vector for a system can be made using the scheme summarized in Figure 4.9. As long as $n_v \geq 3$, component selection and fault isolation is done on the basis of MVS, and coverage is very high (but not perfect). An MVS-based RM scheme can be defeated if a second failure occurs before the first failure is fully processed. Recall that a provisionally failed component is retained for n_f frames, after which point it is removed from $\mathbf{S_v}$, FLC coverage

```
R_OLC[k, p, c]
    n ← length[p]
    P ← array of length[n − k + 1]
    P[1] ← 1
    r ← 0
    for i = 2 upto n − k + 1 do P[i] ← 0 done

    for i = 1 upto n do
      for j = n − k downto 1 do
        if j == n − 1 then P[j+1] ← p_i ⊗ P[j+1] ⊕ (1 − p_i)] ⊗ c ⊗ P[j]
        else P[j + 1] ← p_i ⊗ P[j + 1]] ⊕ (1 − p_i)P[j]
        if i == n then  r ← r ⊕ P[j + 1]
      done
      P[1] ← p_i ⊗ P[1]
    done

    return r ⊕ P[1]
```

Fig. 4.7 Table-based OLC at least k-out-of-n:G algorithm

```
Re_OLC[k, p, c]
    n ← length[p]
    P ← array of length[n − k + 1]
    P[1] ← 1
    for i = 2 upto n − k + 1 do P[i] ← 0 done

    for i = 1 upto n do
      for j = n − k downto 1 do
        if j == n − 1 then P[j+1] ← p_i ⊗ P[j+1] ⊕ (1 − p_i)] ⊗ c ⊗ P[j]
        else P[j + 1] ← p_i ⊗ P[j + 1] ⊕ (1 − p_i) ⊗ P[j]
        if i == n then return P[j + 1]
      done
      P[1] ← p_i ⊗ P[1]
    done

    return P[1]
```

Fig. 4.8 Table-based OLC exactly k-out-of-n:G algorithm

for the first $n - 2$ failures (those occurring while $n_v \geq 3$) can be estimated as the probability that a second like failure will occur during the period of time required to complete the full RM task. If the system runs at a frame (sample) rate Δ_t, then the time required to complete the RM task of identifying and ultimately removing a failed component from $\mathbf{S_v}$ (called the *fault detection window*, w) is

$$w = \Delta_t n_f \ . \tag{4.38}$$

1. Design the system to remain operational after $n - 1$ failures among a set of n redundant elements.
2. Maintain a "voting set," $\mathbf{S_v}$, consisting of the n_v operational components, with $n_v \leq n$.
3. Use a majority voting scheme, as long as $n_v \geq 3$, to select among the n_v components in $\mathbf{S_v}$ (the selected value is v_s). That is, use a mid-value-select (MVS) approach to determine v_s.
4. Use a predetermined "fault detection threshold" (FDT) for each component type such that if $|v_i - v_s| >$ FDT, the ith component is judged to be in a failed state.
5. Declare the ith component "provisionally failed" if $|v_i - v_s| >$ FDT.
6. Declare as failed any component that is provisionally failed for n_f successive frames (samples), and remove it from $\mathbf{S_v}$ (a system with $n_f = 1$ is susceptible to noise-induced faults, but $n_f = 3$ provides a good degree of protection against nuisance failures).
7. Declare a failure for a system with $n_v = 2$ in which the component outputs differ by an amount in excess of the FDT, and make the selection between the two components in disagreement using BIT or some combination of BIT and heuristic tests.

Fig. 4.9 Outline of a redundancy management architecture

The estimation of k-out-of-n:G FLC coverage for a component having a failure rate λ is

$$c_i = e^{-(n-i)\lambda w} . \quad (4.39)$$

Equation (4.39) is the probability that given a failure of a redundant component, a second like component will fail within a period w. Consider a 1-out-of-4:G system; at the time of the first failure, the system is vulnerable to a failure among the remaining three operational components until the RM process has run its course (that is, during the fault detection window, w). The probability that a component with a failure rate λ fails in a period w is $e^{-\lambda w}$, and consequently, the probability that any of the three components will fail is $e^{-3\lambda w}$, a value consistent with Equation (4.39) for the first failure. After the first failure, the system continues to operate as a 1-out-of-3:G system. After the second failure, the system is vulnerable to an additional failure between the remaining two operational components, with the probability of a covered failure being $e^{-2\lambda w}$, a value also consistent with Equation (4.39) for the coverage of a second failure in an FLC 1-out-of-4:G system. Clearly, Equation (4.39) is of use only for estimating coverage for the initial $n-2$ failures. Once the system has failed down to a voting set of two components ($n_v = 2$), coverage is predicated on the probability of success of the BIT or a combination of BIT and heuristic tests. The $(n-1)$th value (the one-on-one coverage value) must be estimated as the probability that the combination of BIT and heuristic tests will be successful.

The following is a *Mathematica* implementation of Equation (4.39):

```
covCal[n_,m_,λ_,w_]:=
Module[{wHr = w/(1000 × 3600)},
N[e^(-(n-m)λwHr), 15]
];
```

where

n Number of redundant elements in the FLC set
m Fault number ($i = 1$ for the first failure, $i = 2$ for the second failure, *etc.*)
λ Redundant element failure rate
w Fault detection window in milliseconds

If λ and **w** are given as exact numbers, **covCal** returns a value accurate to 15 significant digits.

4.9 Comparison of PFC and IFC Systems

This section examines the effects of IFC on redundant systems. Two systems are studied: in this section, a simple 1-out-of-4:G system, and in the next chapter, a more complicated quad-redundant computer control system. It is shown that imperfect fault coverage has a significant effect on the probability of failure for redundant systems and that the character of this effect is fully demonstrated in the simple *k*-out-of-*n*:G system.

One way to demonstrate the effect of imperfect fault coverage on the probability of failure of a redundant system is to examine a series of simple PFC, ELC, FLC and OLC 1-out-of-4:G systems. This section compares, on an equivalent basis, implementations of simple 1-out-of-4:G systems that use RM designed appropriately for ELC and FLC systems. For comparison, the corresponding PFC system and the OLC approximation of the FLC system are shown. Each of these systems consists of four redundant elements, with each element having a constant failure rate $\lambda = 1000$ fpmh. Symbolic results are computed using the algorithms from Section 4.2, although any of the IFC functions presented therein could have been used. These expressions were then used to compute the numerical results shown in Figures 4.10 and 4.11.

For the IFC systems, the redundant elements are each each assumed to have an associated BIT capable of 90% coverage. The systems are discretely operated at a frame rate of 100 Hz, and the selection among the redundant elements is made at this frequency. For the ELC system, the n coverage values are all equal to 0.90. For the FLC system, the $(n-1)$th coverage value, c_3, is also based on BIT coverage and is equal to 0.90. (The coverage value c_3 is the value associated with the last, or $(n-1)$th, failure; after this failure, the system no longer has redundancy.) The RM for the initial FLC system failures is based on a mid-value-select strategy that provides correct selection as long as an additional failure does not occur prior to completion of system reconfiguration. For this analysis, it is assumed that reconfiguration requires three successive frames; at a 100 Hz frame rate, this reconfiguration time is 30 milliseconds. The variable w is defined as the fault detection/reconfiguration window. The FLC coverage values c_1 (which is associated with the first failure coverage) and c_2 (which is associated with the second failure coverage) can thus be estimated using Equation (4.39):

$$c_1 = e^{-(n-1)\lambda w} = 0.999999975 \tag{4.40}$$
$$c_2 = e^{-(n-2)\lambda w} = 0.999999983\,. \tag{4.41}$$

Table 4.3 summarizes the coverage values used for each of the IFC models. For PFC, of course, coverage is not a factor, but the single OLC coverage value, which applies to the third failure, is listed in the c_1 column.

Table 4.3 Coverage values for IFC 1-out-of-4:G models

Model	c_1	c_2	c_3	c_4
PFC	n/a	n/a	n/a	n/a
OLC	0.9	n/a	n/a	n/a
FLC	0.999999975	0.999999983	0.9	n/a
ELC	0.9	0.9	0.9	0.9

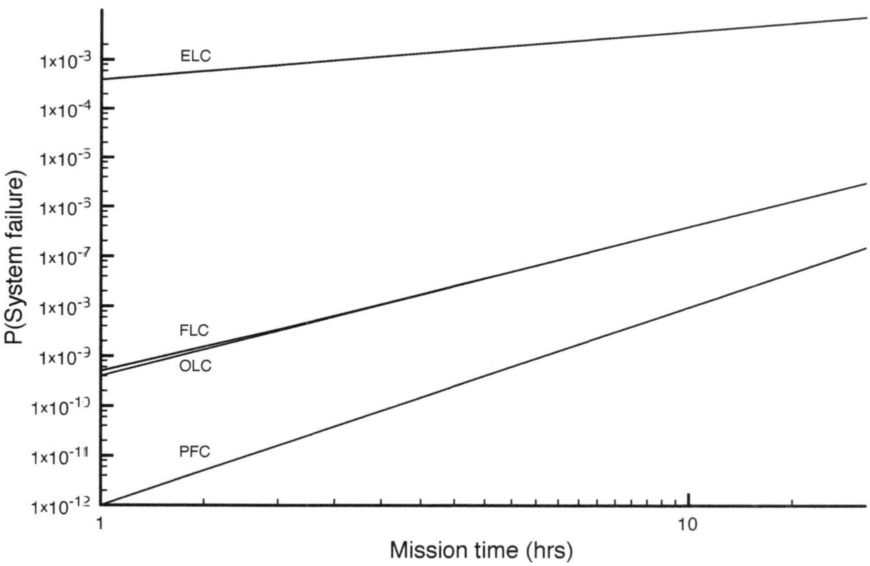

Fig. 4.10 Probability of failure for 1-out-of-4:G systems with BIT coverage value of 0.9

The results, shown graphically as a log-log plot in Figure 4.10 and shown numerically in Table 4.4, clearly demonstrate the importance of correct coverage modeling for the analysis of redundant systems; for mission times less than 10 hours, the FLC and OLC systems are over 40 times more likely to fail than is the PFC system. The reasons for the use of mid-value-select voting-based RM (appropriately modeled

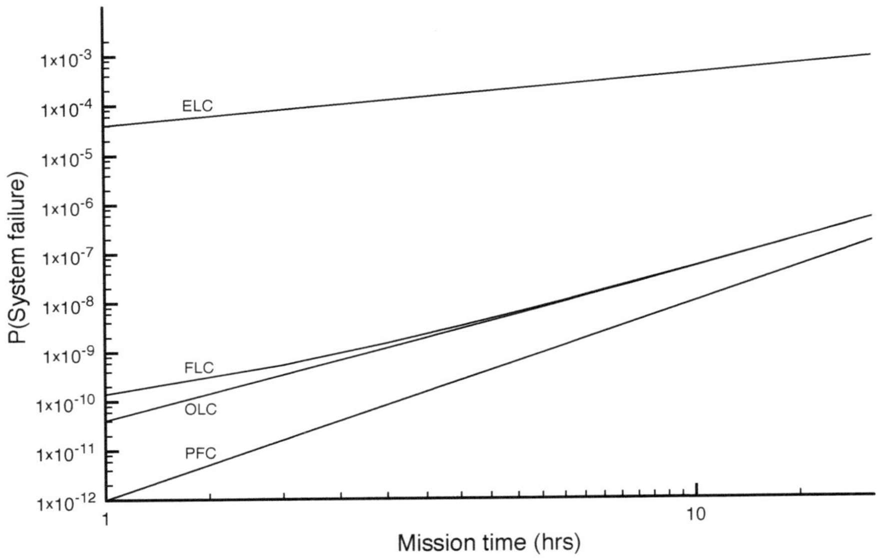

Fig. 4.11 Probability of failure for 1-out-of-4:G systems with BIT coverage value of 0.99

Table 4.4 Probability of failure for 1-out-of-4:G systems with BIT coverage value of 0.9

t (hrs)	PFC	ELC	FLC	OLC
1	9.97424×10^{-13}	3.9974×10^{-4}	4.99899×10^{-10}	3.99999×10^{-10}
2	1.59366×10^{-11}	7.98961×10^{-4}	3.39958×10^{-9}	3.19998×10^{-9}
5	6.18784×10^{-10}	1.99352×10^{-3}	5.04953×10^{-8}	4.99978×10^{-8}
10	9.80215×10^{-9}	3.97414×10^{-3}	4.00921×10^{-7}	3.99931×10^{-7}
20	1.53737×10^{-7}	7.89714×10^{-3}	3.19978×10^{-6}	3.19782×10^{-6}

Table 4.5 Probability of failure for 1-out-of-4:G systems with BIT coverage value of 0.99

t (hrs)	PFC	ELC	FLC	OLC
1	9.97424×10^{-13}	3.99794×10^{-5}	1.40798×10^{-10}	4.08984×10^{-11}
2	1.59366×10^{-11}	7.99177×10^{-5}	5.33941×10^{-10}	3.3434×10^{-10}
5	6.18784×10^{-10}	1.99487×10^{-4}	6.05419×10^{-9}	5.55669×10^{-9}
10	9.80215×10^{-9}	3.97957×10^{-4}	4.98051×10^{-8}	4.8815×10^{-8}
20	1.53737×10^{-7}	7.91966×10^{-4}	4.60105×10^{-7}	4.58145×10^{-7}

with FLC), rather than a BIT-based approach (modeled with ELC), by systems that require extremely low probabilities of failure also become apparent. As would be expected, the PFC system has a substantially lower probability of failure than any of the IFC systems. The PFC curve has a slope of approximately 4 dpd; that is, the slope measured in decades per decade on a log-log plot is approximately equal to the level of redundancy, n, in the system.

The OLC approximation in this case is excellent for mission times of approximately 3 hours or greater. For mission times greater than 3 hours, both the OLC and FLC curves have a slope of approximately 3 dpd (that is, $n - 1$). It is a general relationship that PFC systems have a slope of approximately n, and FLC and OLC systems have a slope of approximately $n - 1$. Thus, the slope of the probability of system failure curve is a measure of its equivalent redundancy level. In effect, then, the presence of less-than-perfect coverage has a "cost" of approximately one level of redundancy. The ELC system has a slope of approximately 1 dpd—the same as that of a simplex system. The curve for a simplex system, also with $\lambda = 1000$ fpmh, has the same slope but is translated vertically.

Figure 4.11 and Table 4.5 show the results of repeating the experiment with the BIT-based coverage increased from 90% to 99%. Even with this relatively high level of BIT coverage, the same general observations regarding the effects of coverage still hold.

A comparison of Figures 4.10 and 4.11 shows that increasing the level of BIT coverage from 0.9 to 0.99 causes the effective range of the OLC approximation of FLC to increase from approximately 2 hours to approximately 4 hours. Even with the increase in BIT coverage to a relatively high level of 0.99, FLC systems still far outperform ELC systems, since the ELC systems are nearly eight thousand times more likely to fail for a mission time of 10 hours. It is also apparent that the assumption that using mid-value-select voting results in perfect coverage (thus supporting the use of OLC) is not always valid. In this particular circumstance, however, such an assumption would be valid if the system reliability requirements were for mission times greater than 3 hours.

Although these results are for a simple 1-out-of-4:G system, it is shown in the following chapter that the general characteristics of a multichannel redundant system are the same. These results also clearly demonstrate the reasons why highly reliable systems use RM-based MVS voting. That is, the results demonstrate why these highly reliable systems are FLC systems instead of ELC systems that rely solely on BIT as their primary RM technique.

References

1. Myers AF (2007) k-out-of-n:G System Reliability With Imperfect Fault Coverage. IEEE Trans Relia 56:464–473
2. Myers A, Rauzy A (2008) Efficient Reliability Assessment of Redundant Systems Subject to Imperfect Fault Coverage Using Binary Decision Diagrams. IEEE Trans Relia 57:336–348
3. Myers AF, Rauzy A (2008) Assessment of redundant systems with imperfect coverage by means of binary decision diagrams. Reliab Eng Syst Saf 93:1025–1035

Chapter 5
Complex System Modeling Using BSV

"What's the problem?" you say?
'Tis this.

Abstract Chapter 3 demonstrated the use of the Bernoulli state variable (BSV) technique in the assessment of systems with complex interconnections. This chapter illustrates the use of BSV in the reliability assessment of more complex systems consisting of multiple k-out-of-n:G blocks with both perfect and imperfect fault coverage. Here, the redundant inputs to a k-out-of-n:G selector can be represented by complex polynomials as opposed to just a set of disjoint literals.

5.1 Background

This chapter demonstrates the use of the BSV technique, implemented using *Mathematica*, for the assessment of redundant systems that are more complex than the simple k-out-of-n:G systems discussed in previous chapters. The BSV technique has been previously presented in [1], which also assesses a version of the computer control system covered in Section 5.3.

The modeling of imperfect fault coverage, to the extent that it has been addressed in the literature, has typically been treated as the selection among single sets of redundant components (for example, see [2–8]. Nevertheless, the effects of imperfect fault coverage often must be applied to reliability polynomials that represent "upstream" components and subsystems. In computer-controlled systems with redundancy management that must select among redundant inputs, the redundant inputs often represent the output of upstream components that may also have complex dependencies. This situation frequently results in non-disjoint inputs. For the systems that are analyzed in this chapter, the effects of imperfect fault coverage are applied to upstream polynomials, as opposed to single sets of redundant components. These upstream polynomials may or may not be disjoint.

Many of the systems analyzed in this book are made up of a number of redundant subsystems, each with the same level of redundancy. It is useful to refer to each individual redundancy level (set of redundant components) as a *channel*; that is, if the overall system or a major portion thereof has a redundancy level of three, then the system is said to have three channels. It is also useful to refer to components within the same channel as being *local* components and to refer to components of the same type but in different channels as being *sister* components.

5.2 Blocks of Redundant Components in Series

This section covers the reliability modeling of systems made up of multiple sets of redundant components configured in series. In a subsystem having redundant components, there must exist some means for the subsystem block to accomplish the RM tasks of fault detection, isolation and accommodation in the event of a failure among the redundant components in the block. These RM tasks can be accomplished in various ways. One technique involves the use of an external RM device;[1] an example of this arrangement is shown in Figure 5.1. The system depicted in the figure could be a PFC, ELC, FLC or OLC system or subsystem, depending on the architecture of the RM implemented in the block labeled R.

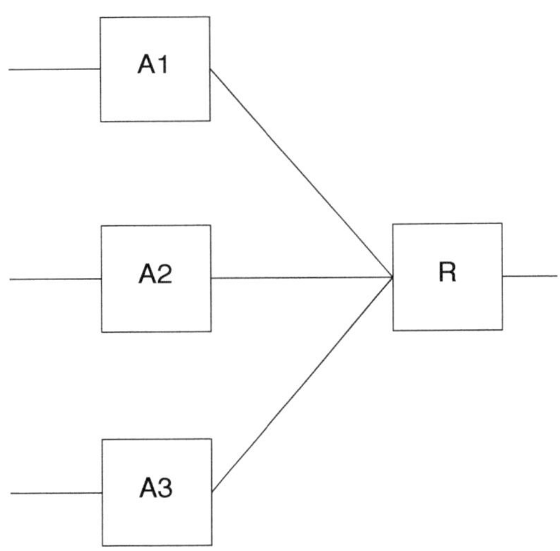

Fig. 5.1 Redundant system with external RM

[1] If the redundant set is implemented as an FLC or OLC system, this external RM device is frequently referred to as a *voter*, since the principle element of the RM scheme would use some form of mid-value-select voting.

The RM function can also be incorporated within the redundant set, **A**, Accomplishing this task, however, requires that each of the redundant components be interconnected, such as is shown in Figure 5.2.

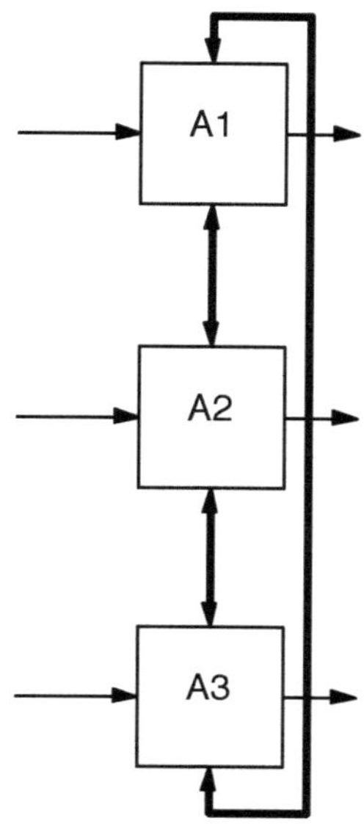

Fig. 5.2 Redundant system with internal RM

If the redundant components A_i in Figure 5.2 are digital devices, then the interchannel paths are frequently called a *cross-channel data link* (CCDL). This system is able to accomplish its RM internally because each of the A_i components has knowledge of the input, output and status of both itself and each of its sister components. Depending on how the RM tasks are accomplished and on whether the RM design uses voting or component health monitoring (possibly BIT), the system shown in Figure 5.2 could be either an FLC, OLC or ELC system. If the RM task could be done with perfect certainty, then the system would be a PFC system.

Redundant k-out-of-n:G systems can be structured with either an internally or externally implemented RM capability; in subsequent sections, both of these structures are encountered. Consider the collection of components shown in Figure 5.3, which depicts three blocks containing redundant components: Block A, Block B and Block C. The third block, Block C, consists of a simplex component. The reliability of an overall system comprising the components included in these three blocks depends both on how the blocks themselves are interconnected and on how the components within the blocks are interconnected (that is, whether they use internal or external RM).

Obviously, the components shown in Figure 5.3 could be interconnected in a large number of ways. The possible interconnections that effectively use the redun-

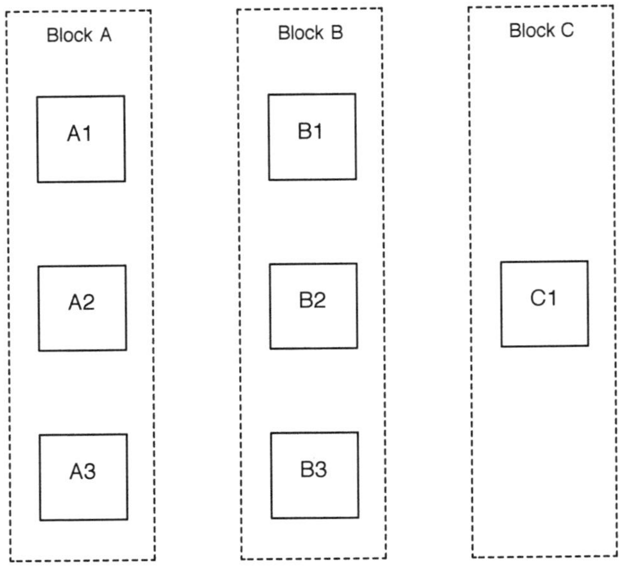

Fig. 5.3 Redundant components

dant components can be reduced, however, to two configurations, which are designated Configuration 1 and Configuration 2.

5.2.1 Configuration 1

Figure 5.4 illustrates one possible arrangement of the components that are shown in Figure 5.3. In this arrangement, both the Block A and Block B components are interconnected using CCDLs, and both use internal RM. As shown in Figure 5.4, the output of Block A is transmitted to each of the components in Block B. If the RM implemented in Block A not only uses the CCDL to share the status of the A_i components and their inputs[2] but also uses the CCDL to share the RM selection made by each A_i with each of its sister components,[3] then this configuration can be equivalently implemented as shown in Figure 5.5. In this implementation, each A_i communicates only with its local B_i. The two above-mentioned configurations are equivalent because, in both cases, the RM for Block A is performed internally, and the result is communicated to each of the B_i components; these two configurations, 1a and 1b, are collectively referred to as Configuration 1.

System reliability for Configuration 1 can easily be derived using BSV. Modeling these systems involves the k-out-of-n:G selection among redundant subsystem

[2] In an FLC or OLC system, this is often referred to as an *input voting plane*.

[3] This is often referred to as an *output voting plane*.

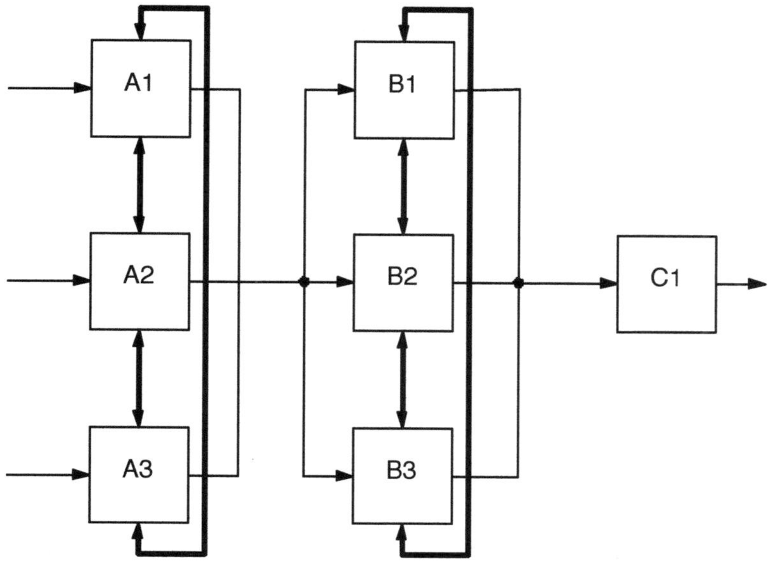

Fig. 5.4 Redundant system—Configuration 1a

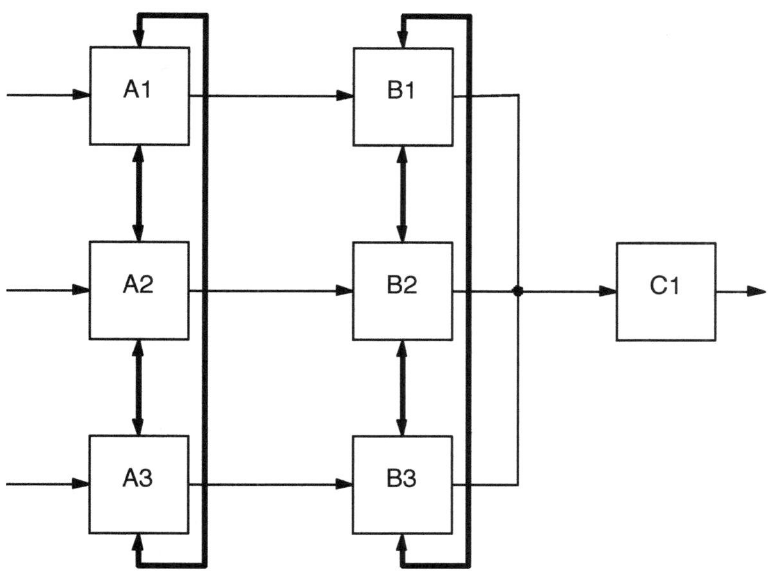

Fig. 5.5 Redundant system—Configuration 1b

elements that are not necessarily disjoint. The BSV implementation of the ⊗ and ⊕ operators, along with the **RaL** function described in Chapter 4, can be used to determine the reliability of systems using ELC, FLC, OLC or PFC models. Shown below is the *Mathematica* code that implements a BSV derivation of a PFC model for Configuration 1.

Define a conversion rule to convert the result to a function in **r[λ, t]:**

```
TOr = {A1 → r[λA, t], A2 → r[λA, t], A3 → r[λA, t],
B1 → r[λB, t], B2 → r[λB, t], B3 → r[λB, t],
C1 → r[λC, t]};
```

The following is the Configuration 1 model:

```
sA = {A1, A2, A3};
sB = {B1, B2, B3};
sAvote = RaL[1, sA];
sBvote = RaL[1, sB];
sSysConf1 = sAvote ⊗ sBvote ⊗ C1;
rSysConf1 = sSysConf1/.TOr
```

$$9\, r[\lambda A,t]\, r[\lambda B,t]\, r[\lambda C,t] - 9\, r[\lambda A,t]^2\, r[\lambda B,t]\, r[\lambda C,t] + 3\, r[\lambda A,t]^3\, r[\lambda B,t]\, r[\lambda C,t] - 9\, r[\lambda A,t]\, r[\lambda B,t]^2\, r[\lambda C,t] + 9\, r[\lambda A,t]^2\, r[\lambda B,t]^2\, r[\lambda C,t] - 3\, r[\lambda A,t]^3\, r[\lambda B,t]^2\, r[\lambda C,t] + 3\, r[\lambda A,t]\, r[\lambda B,t]^3\, r[\lambda C,t] - 3\, r[\lambda A,t]^2\, r[\lambda B,t]^3\, r[\lambda C,t] + r[\lambda A,t]^3\, r[\lambda B,t]^3\, r[\lambda C,t]$$

This is a straightforward BSV model of two 1-out-of-3:G subsystems in series with the simplex element C_1.

5.2.2 Configuration 2

A second way in which the components in Figure 5.3 can be interconnected is shown in Figure 5.6. For Configuration 2, each B_i component is dependent on its sister B_i component, which means that a Block B component cannot contribute to the overall system reliability unless its corresponding Block A component is also operational. Also notice that in Configuration 2, Block C must still play a role in the overall system RM task since it must have some means of selecting among the redundant Block B outputs. It should also be clear that Configuration 2 has fewer success paths than Configuration 1 and that Configuration 2 consequently has a lower reliability.

The modeling of Configuration 2 must incorporate the dependence of each individual B_i component on its local A_i component. This incorporation is the only modification that is made to the code used above for Configuration 1, other than the change in the variable names for the result.

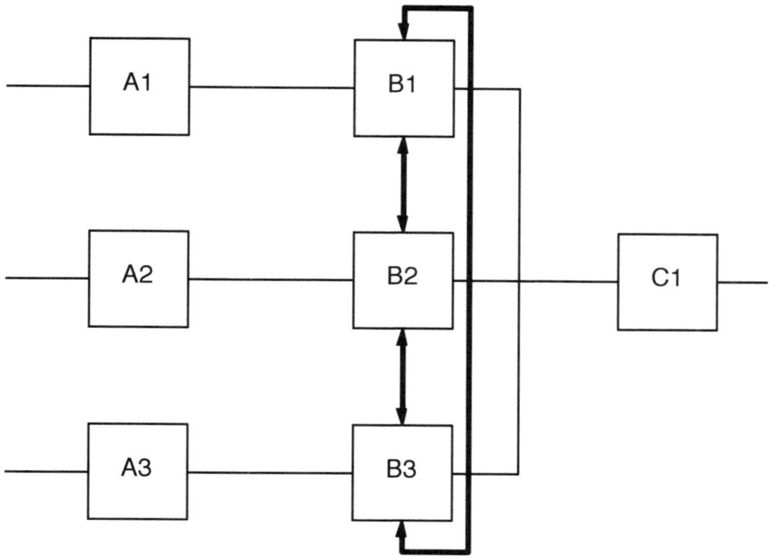

Fig. 5.6 Redundant system—Configuration 2

```
sA = {A1, A2, A3};
sB = {B1, B2, B3};
sAvote = RaL[1, sA];
sBvote = RaL[1, sA ⊗ sB];
sSysConf2 = sAvote ⊗ sBvote ⊗ C1;
rSysConf2 = sSysConf2/.TOr
```

$$3\, r[\lambda A, t]r[\lambda B, t]r[\lambda C, t] - 3\, r[\lambda A, t]^2 r[\lambda B, t]^2 r[\lambda C, t] + r[\lambda A, t]^3 r[\lambda B, t]^3 r[\lambda C, t]$$

The reliability polynomials for Configurations 1 and 2 are clearly different, and as discussed above, Configuration 1 is expected to have a reliability superior to that of Configuration 2.

5.2.3 Comparison of Configurations 1 and 2

In both configurations, the redundant component sets, Blocks A and B, are in series with a simplex Block C component. Unless the simplex component, C_1, has a failure rate λ_C that is much less than both λ_A and λ_B, the C_1 reliability will dominate the overall system reliability. The effect of the structural differences between Configurations 1 and 2 can be seen in Figure 5.7 with $\lambda_A = \lambda_B = 500$ fpmh and $\lambda_C = 1/1000$ fpmh.

The following code produces a table of values for the probability of failure for Configuration 1 and Configuration 2. The calculation of the numerical results to six significant digits[4] requires the determination of the FLC coverage values to higher-than-normal machine-level precision. The following version of the **covCal** function provides 25 digits, as opposed to the normal machine-level precision of 16–18 significant digits.

```
covCal[n_,m_,λ_,w_]:=
Module[{wHr = w/(1000×3600)}, N[e^(-(n-m)λwHr), 25]]
```

The **setNumeric** and **setSymbolic** routines shown below can be used to easily switch between symbolic and numerical calculations:

```
setNumeric:={
λA = 500 × 10^-6;
λB = 500 × 10^-6;
λC = 1/1000 10^-6;
r[λ_,t_]:=e^(-λt);
};

setSymbolic:=Clear[λA, λB, λC, r, t];
```

The probabilities of failure for Configurations 1 and 2 are computed below.

```
setNumeric;
times = {1,2,3,10,20};
Table[{t = times[[i]], 1 - rSysConf1//N,
1 - rSysConf2//N}, {i, 1, Length[times]}]//TableForm
setSymbolic;
```

1	1.24981^{-9}	1.99850^{-9}
2	3.9970^{-9}	9.97604^{-9}
5	3.61331^{-8}	1.29066^{-7}
10	2.58133^{-7}	9.95124^{-7}
20	1.99025^{-6}	7.78395^{-6}

The failure probabilities for Configurations 1 and 2 are shown graphically in Figure 5.7. Even though the component C_1 has a very low failure rate ($\lambda_C = 1/1000$ fpmh), the system reliability is still heavily influenced by this simplex component for mission times less than 1 hour, where the slope of the curve approaches 1 dpd. For greater mission times, both the Configuration 1 and Configuration 2

[4] The reporting of the probability of system failure to a precision of six significant digits cannot be justified, of course, since these results are dependent on failure rate and coverage estimates that have only one or two significant digits and that are unlikely to be known in the real world to any higher precision. The results are expressed with six significant digits only for the purpose of providing an arithmetic check of the underlying algorithmic approach and *not* because the resulting determination of the failure probabilities are justified to this level of precision.

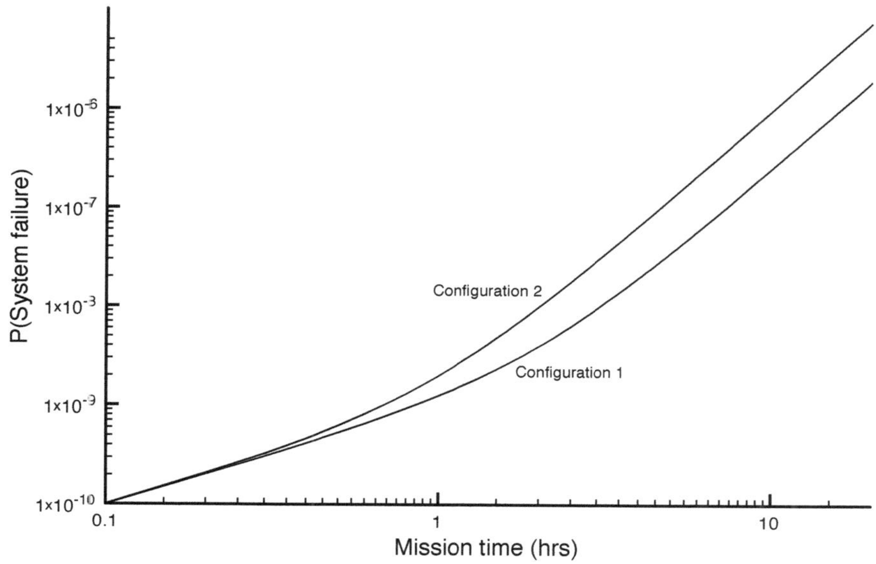

Fig. 5.7 Probability of system failure for Configurations 1 and 2

curves approach a slope of 3 dpd, as would be expected for a PFC triplex system. Also notice that the Configuration 1 system reliability exceeds the Configuration 2 system reliability and that at a mission time of 20 hours, Configuration 2 is nearly four times more likely to fail than is Configuration 1.

This section has demonstrated the use of BSV to analyze two different configurations of redundant PFC subsystems—Blocks A and B—arranged in series with a simplex element—Block C. If these were IFC systems, the derivation would be performed in the same fashion but would use the **RaL** functions for the ELC, FLC or OLC case as appropriate.

The examples analyzed in this section demonstrate the importance of paying careful attention to the details of system configuration and of understanding the details of system RM architecture in the modeling of system reliability. Redundant system reliability cannot be modeled correctly without detailed knowledge of where and how the RM tasks are accomplished.

In this section, it has been shown that redundant system reliability is dependent not only on the redundancy level, but also on the details of the interconnections and RM architecture. The significance of the differences between Configurations 1 and 2 are exploited in the development of the conditional probability models analyzed in Chapter 6.

5.3 Quadruplex Computer Control System

Figure 5.8 shows the functional block diagram of a portion of a simple hypothetical quadruple-redundant real-time control system. The system is operational if and only if it contains at least one operational path.

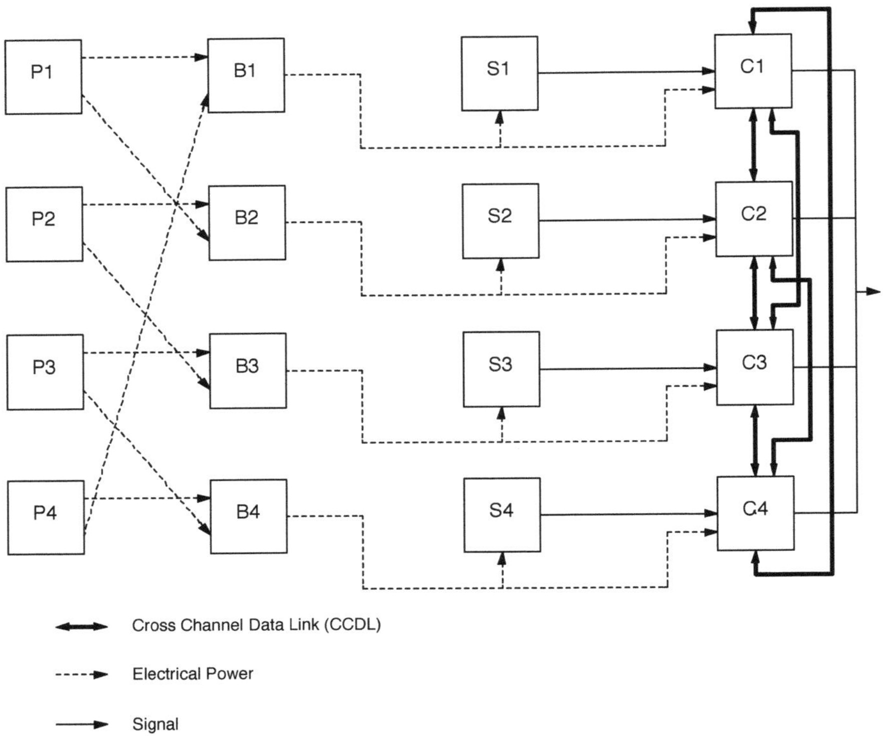

Fig. 5.8 Simple quadruplex computer control system

The quadruple-redundant element sets for this system include **P,** a set of independent electrical power sources, and **B,** a set of electrical power distribution buses. Each bus receives power from two power sources, and since they contain no active elements, the reliability of the buses is completely defined by the power sources that supply them. The other two quadruple-redundant element sets are **S,** a set of feedback sensors that are powered by their corresponding buses (the output from each sensor is an input to the corresponding local control computer), and **C,** a set of control computers, which are also powered by their local buses. The selected output of the computer set provides command signals to a downstream actuation system, which is not included in this model.

Each of the computers in the set communicates with the other three computers through bidirectional CCDLs. The sensors and computers both have BIT capability.

The CCDL provides the local sensor output data and sensor BIT status from each computer to each of the other computers. Each computer then uses the data provided by the CCDL, along with its local data, to select a sensor output value. The computer combines this selected sensor value with a set of system control laws to compute the actuator system commands. These commands are then shared with the other three computers through the CCDL. Each computer then selects a set of command values from the commands that are computed by each operational computer and these command values are transmitted to the actuation system.

The unattended (without repair) reliability of the system shown in Figure 5.8 is equal to the probability that at least one of the four channels is operational. The *Mathematica* calculation in the following section symbolically computes the reliability of an FLC version of this system using the **RaL** function, as defined in Section 2.8.3, and using the BSV implementation of the $\oplus$ and $\otimes$ operators. This represents a situation in which the reliability of the redundant inputs to **C** are themselves reliability polynomials as opposed to simple component reliabilities found in a simple k-out-of-n:G structure. The results for PFC, ELC and OLC systems can be calculated in a similar fashion by making the appropriate changes to accommodate the differences in coverage values that are required for each of the models. The resulting expressions are used to compute the reliability for each system.

Table 5.1 Quadruple-redundant system characteristic values

Computer frame rate	100 Hz (10 ms)
Fault detection window, w	3 frames = 30 ms
Power failure rate, λ_P	200 fpmh
Sensor failure rate, λ_S	250 fpmh
Computer failure rate, λ_C	400 fpmh
Sensor BIT coverage	0.9
Computer BIT coverage	0.95

Table 5.2 Sensor and computer coverage values

Model	Element	c_1	c_2	c_3
PFC	Sensor	–	–	–
	Computer	–	–	–
OLC	Sensor	`cSolc = 0.9`	–	–
	Computer	`cColc = 0.95`	–	–
FLC	Sensor	`cS1 = 0.99999999375`	`cS2 = 0.99999999583`	`cS3 = 0.9`
	Computer	`cC1 = 0.9999999900`	`cC2 = 0.9999999933`	`cC3 = 0.95`

The element BIT coverage values and other system characteristics are listed in Table 5.1. For PFC, no coverage values are required. For the OLC model, the one-on-one coverage is estimated as the BIT coverage of the corresponding element. For the FLC model, the first and second failure coverage values for the sensors and computers are estimated as the conditional probability that an additional failure of the same element type will occur within the fault detection window, w. This approach is the same as the one outlined in Section 4.8 and implemented using Equation (4.41). These computed coverage values are included in Table 5.2.

5.3.1 Mathematica *Code for the Simple Quadruplex Computer Control System*

The following *Mathematica* code uses the previously defined functions to compute the probability of failure of the simple quadruplex computer control system depicted in Figure 5.8. This code exemplifies how a system model is built, given a functional block diagram description of the system. The code below derives a model for an FLC implementation of the system. Following the derivation of the FLC implementation, both PFC and OLC expressions are obtained as well.

Define the problem variables:

```
sP = {P1, P2, P3, P4};
sS = {S1, S2, S3, S4};
sC = {C1, C2, C3, C4};
```

Define the FLC vectors for the sensors and computers:

```
cS = {cS1, cS2, cS3};
cC = {cC1, cC2, cC3};
```

Compute the bus state:

```
sBout = {P1 ⊕ P4, P2 ⊕ P1, P3 ⊕ P2, P4 ⊕ P3};
```

Both the computers, **sC,** and the sensors, **sS,** are dependent on the bus state (that is, they need power):

```
sSpwr = sS ⊗ sBout;
sCpwr = sC ⊗ sBout;
```

For a sensor to be included in the vote, its local computer must be operational; the second argument of the **RaL** function is then **sSpwr ⊗ sC.** Compute the result of the sensor vote:

```
sSvoted = RaL[1, sSpwr ⊗ sC, cS];
```

Calculate the result of the computer vote and then the overall system reliability:

```
sCvoted = RaL[1, sCpwr, cC];
sSys = sSvoted ⊗ sCvoted;
```

The term **sSvoted** represents the output of the input voting plane. The reliability of the overall computer system depends on both of these votes, and **sSys** represents the output of the output voting plane, which is the overall reliability of the system shown in Figure 5.8. The resulting reliability polynomial has 912 terms:

```
sSys//Length
912
```

Convert the results to a polynomial in **r[λ, t]** using the following substitution rule:

```
TOr = {
P1 → r[λP, t], P2 → r[λP, t], P3 → r[λP, t], P4 → r[λP, t],
S1 → r[λS, t], S2 → r[λS, t], S3 → r[λS, t], S4 → r[λS, t],
C1 → r[λC, t], C2 → r[λC, t], C3 → r[λC, t], C4 → r[λC, t]
};

rSysFLC = sSys/.TOr;
```

The polynomial in **r[λ, t]** has 78 terms:

```
rSysFLC//Length
78
```

By expressing the polynomial in terms of i.i.d. components and as a function of **r[λ, t]**, the size of the polynomial is reduced from 912 to 78 terms. The first and last terms of the polynomial are

```
rSysFLC//First
8 cC1 cC2 cC3 cS1 cS2 cS3 r[λC, t]r[λP, t]r[λS, t]

rSysFLC//Last
4 cS1 cS2 cS3 r[λC, t]^4 r[λP, t]^4 r[λS, t]^4
```

When dealing with fully expanded polynomials that represent the reliability of a system, it is always useful to examine at least the first and last polynomial terms. The first term always represents the components that constitute a minimum path required for the system to be operational. In this case, for example, the functions **r[λC, t]**, **r[λP, t]** and **r[λS, t]** included in the first polynomial term are all to the first power. Since, as a minimum, at least one of each of these components is required for the system to be operational, their product represents a minimum path for the system. The last term always includes expressions for the maximum number of components that are possible in the system. In this case, each of the functions is raised to the fourth power. These relationships always exist in a fully expanded reliability polynomial, and checking for them is a good (but not sufficient) method for verifying the validity of the expression. If either the first term or the last term of the fully expanded polynomial fails to meet these requirements, the formulation contains an error.

During the analysis of system reliability, the ability to readily convert from a symbolic representation to a numerical one is useful; the following routines, **setNumeric** and **setSymbolic,** provide this capability.

```
setNumeric:=(

λP = 200 × 10^-6;
λS = 250 × 10^-6;
λC = 400 × 10^-6;

cSolc = 9/10;
cColc = 95/100;

cS1 = covCal[4, 1, λS, 30];
cS2 = covCal[4, 2, λS, 30];
cS3 = cSolc;

cC1 = covCal[4, 1, λC, 30];
cC2 = covCal[4, 2, λC, 30];
cC3 = cColc;

r[λ_, t_]:=e^(-λt);
);

setSymbolic:=Clear[λP, λS, λC, cS, cC, , cSolc, cColc, cS1, cS2, cS3, cC1, cC2,
cC3, r, t];
```

The **setNumeric** routine can be used to facilitate the calculation of a table for the probability of failure of the computer system. Once the numerical calculations are complete, the call to **setSymbolic** restores **rSysFLC** to a symbolic polynomial.

```
setNumeric;
times = {1, 2, 5, 10, 20}; Table[{t = times[[i]], N[1 - rSysFLC, 6]},
{i, 1, Length[times]}]//TableForm
setSymbolic;
```

1	1.53622×10^{-10}
2	1.03566×10^{-9}
5	1.53420×10^{-8}
10	1.21845×10^{-7}
20	9.73925×10^{-7}

An expression for a PFC model of this system can be obtained from the FLC model by substitution. A PFC system is equivalent to an FLC system with all of the coverage values set to unity.

```
rSysPFC = rSysFLC/.cS1 → 1/.cS2 → 1/.cS3 → 1/.cC1 → 1/.cC2 →
1/.cC3 → 1;
```

Compared with the 78 terms in the FLC polynomial, the PFC polynomial has only 10 terms.

```
rSysPFC//Length
```
10

Again, the first and last terms can be obtained:

```
rSysPFC//First
```
$8\, r[\lambda C, t] r[\lambda P, t] r[\lambda S, t]$

```
rSysPFC//Last
```
$r[\lambda C, t]^4 r[\lambda P, t]^4 r[\lambda S, t]^4$

A table of values for the probability of system failure for the PFC computer system is calculated in the same fashion:

```
setNumeric;
times = {1,2,5,10,20}; Table[{t = times[[i]], N[1 - rSysPFC,6]},
{i,1,Length[times]}]//TableForm
setSymbolic;
```

1	1.79931×10^{-13}
2	2.87610×10^{-12}
5	1.12020×10^{-10}
10	1.78359×10^{-9}
20	2.82595×10^{-8}

An expression for an OLC model of the system is derived in a similar fashion; an OLC system is equivalent to an FLC system with the initial coverage values set to unity:

```
rSysOLC =
  rSysFLC/.cS1 → 1/.cS2 → 1/.cS3 → cSolc/.cC1 → 1/.cC2 →
1/.cC3 → cColc;
```

```
rSysOLC//Length
```
34

```
rSysOLC//First
```

$8\,\text{cColc cSolc}\, r[\lambda\text{C}, t] r[\lambda\text{P}, t] r[\lambda\text{S}, t]$

```
rSysOLC//Last
```
$4\,\text{cSolc}\, r[\lambda\text{C}, t]^4 r[\lambda\text{P}, t]^4 r[\lambda\text{S}, t]^4$

A table of values for the probability of system failure for the OLC computer system can be calculated as follows:

```
setNumeric;
times = {1,2,5,10,20}; Table[{t = times[[i]], N[1 - rSysOLC, 6]},
{i, 1, Length[times]}]//TableForm
setSymbolic;
```
1 1.21386×10^{-10}
2 9.71217×10^{-10}
5 1.51811×10^{-8}
10 1.21524×10^{-7}
20 9.73286×10^{-7}

In this section, PFC and OLC expressions were derived from the FLC expression. The techniques used in these derivations are general and can be applied in any similar circumstance. The PFC and OLC expressions could, of course, have been computed directly in a fashion similar to that used for the FLC expression.

5.3.2 Quadruplex Computer System Results

The probability of failure for the quadruplex computer control system is shown in Table 5.3. The symbolic expressions for the PFC, FLC and OLC versions of the system are used to compute the probability of system failure, which is shown graphically in Figure 5.9.

Table 5.3 Quadruplex computer system probability of failure

t (hrs)	PFC	FLC	OLC
1	1.79931×10^{-13}	1.53622×10^{-10}	1.21386×10^{-10}
2	2.87610×10^{-12}	1.03566×10^{-9}	9.71217×10^{-10}
5	1.12020×10^{-10}	1.53420×10^{-8}	1.51811×10^{-8}
10	1.78359×10^{-9}	1.21845×10^{-7}	1.21524×10^{-7}
20	2.82595×10^{-8}	9.73925×10^{-7}	9.73286×10^{-7}

For a mission time of 1 hour, the FLC system is two orders of magnitude more likely to fail than is the PFC system, clearly demonstrating the importance of correctly modeling the effects of IFC. The PFC model has a slope of approximately 4 dpd, and the OLC model has a slope of approximately 3 dpd; these results are as

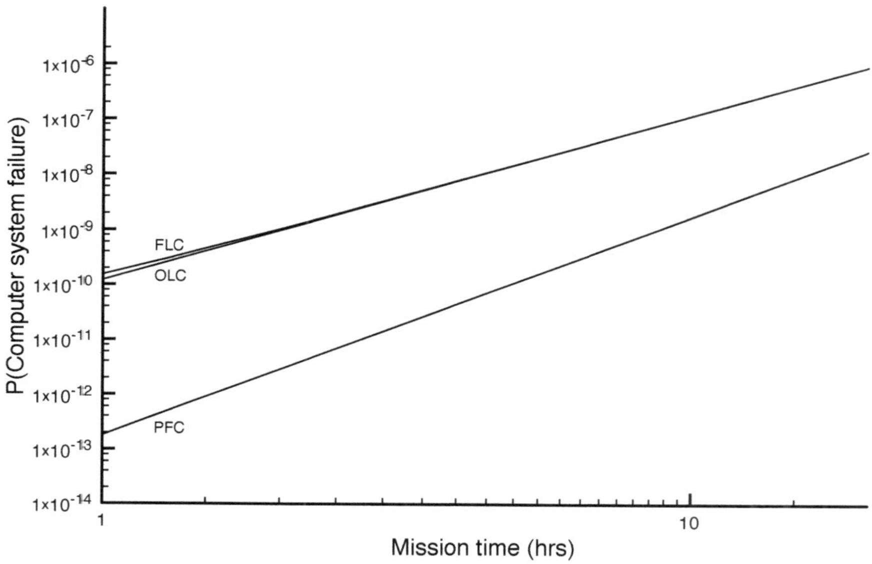

Fig. 5.9 Quadruplex computer system probability of failure

expected for PFC and OLC quadruplex systems. The OLC model is an excellent approximation of the more rigorous and complex FLC model for mission times greater than a few hours.

5.4 Actuation Subsystem

Figure 5.10 depicts a modification of the computer control system functional block diagram to include the interface between the computer control system and the actuation system shown in Figure 5.11. Values for the failure rates and coverage values of the actuation subsystem are given in Table 5.4.

Table 5.4 Actuation system characteristic values

Computer frame rate	100 Hz (10 ms)
Fault detection window, w	3 frames = 30 ms
Actuation computer failure rate, λ_{AC}	400 fpmh
Servo-electronics failure rates, λ_{EL}, λ_{ER}	10 fpmh
Actuator failure rate, λ_A	2 fpmh
AC BIT coverage	0.95
Servo BIT coverage	0.9

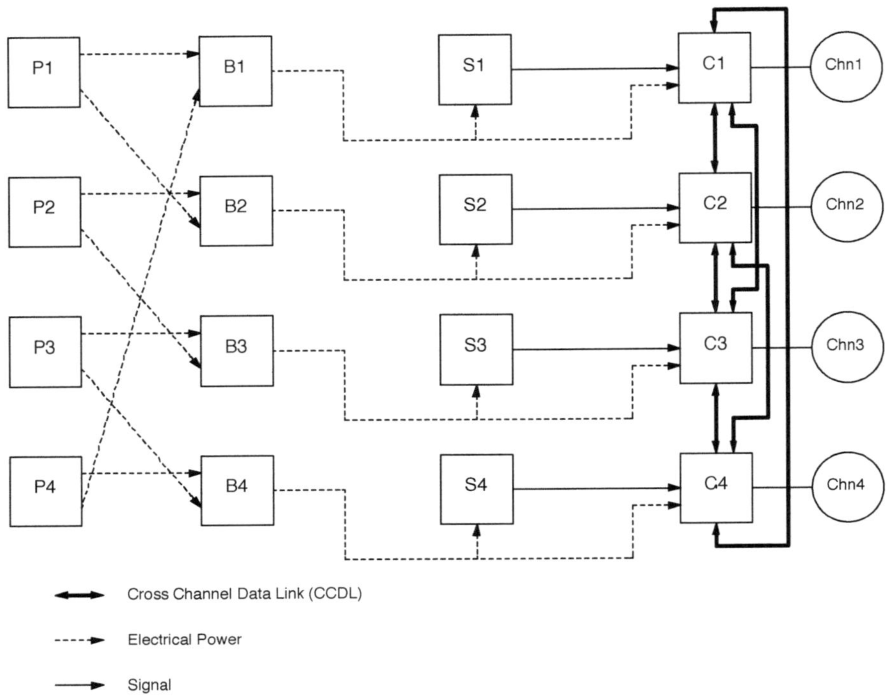

Fig. 5.10 Simple quadruplex system with actuation interface

The quadruplex computer control system shown in Figure 5.8 has been redrawn in Figure 5.10 to include an output ($Chn_1, \ldots, Chn_4$) from each of the computers. Each of the redundant channels has an interface with the actuation system. The actuation system, shown in Figure 5.11, consists of a quadruple set of actuation computers, $AC_1, \ldots, AC_4$, each of which provides commands to two sets of servo-loop electronics: a left side, EL_i, and a right side, ER_i. Each set of servo-loop electronics controls an actuator, A_i. In this system, at least one of the two actuators A_i must be operational for the system to be operational. The characteristic values for the actuation system are given in Table 5.4. The actuators are electrically powered[5] and can be bypassed in the event of a failure.[6] Although the links in Figure 5.11 show only the command path, they are also able to communicate actuator status to the actuation system computers. In the event of an actuator jam, these computers disengage the jammed actuator. The servo commands from the quad servo-electronics, EL_i and ER_i, are "force-summed" on a common armature. Also, the system uses

[5] To keep this example simple, the dependence of the actuator on electrical power has not been included in the analysis.

[6] The ability to bypass a failed electrical actuator is problematic, but the capability is hypothesized for this example.

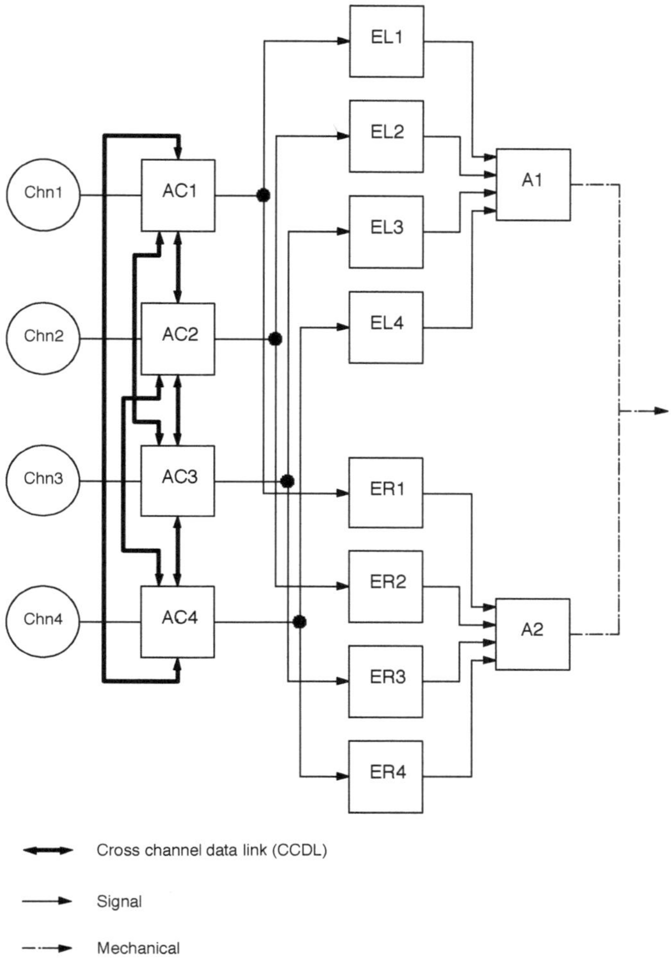

Fig. 5.11 Simple quadruplex actuation subsystem

an electrical drive current model that has a 90% (servo BIT coverage) probability of detecting a failure in the servo command path and of subsequently disabling its commands.

5.4.1 Mathematica *Code for Actuation Subsystem*

The *Mathematica* model shown below (which is derived using BSV) is a straightforward implementation that follows directly from the functional block diagram in

Figure 5.11. The FLC model for the actuation system is developed in the same fashion as is the computer control subsystem model discussed in the preceding section.

```
sAC = {AC1, AC2, AC3, AC4};
sEL = {EL1, EL2, EL3, EL4};
sER = {ER1, ER2, ER3, ER4};

cAC = {cAC1, cAC2, cAC3};
cEL = {cEL1, cEL2, cEL3};

sACvoted = RaL[1, sAC, cAC];
sELselect = sACvoted ⊗ Ral[1, sEL ⊗ sAC, cEL];
sERselect = sACvoted ⊗ Ral[1, sER ⊗ sAC, cER];
sACT1out = A1 ⊗ sELselect;
sACT2out = A2 ⊗ sERselect;
sSys = sACT1out ⊕ sACT2out;

sSys//Length
2912
```

The substitution rule for converting to a polynomial in $r[\lambda, t]$ also includes rules for converting both the **cELn** and **cERn** to **cEn.**

```
TOr = {AC1 → r[λAC, t], AC2 → r[λAC, t], AC3 → r[λAC, t], AC4 → r[λAC, t],
EL1 → r[λE, t], EL2 → r[λE, t], EL3 → r[λE, t], EL4 → r[λE, t],
ER1 → r[λE, t], ER2 → r[λE, t], ER3 → r[λE, t], ER4 → r[λE, t],
cEL1 → cE1, cEL2 → cE2, cEL3 → cE3,
cER1 → cE1, cER2 → cE2, cER3 → cE3,
A1 → r[λA, t], A2 → r[λA, t]};
rSysFLC = sSys/.TOr;

rSysFLC//Length
133

rSysFLC//First
8 cAC1 cAC2 cAC3 cE1 cE2 cE3 r[λA, t]r[λAC, t]r[λE, t]

rSysFLC//Last
-16 cE1^2cE2^2cE3 ^2r[λA, t]^2r[λAC, t]^4r[λE, t]^8
```

Routines for switching between symbolic and numerical calculations are given below:

```
setNumeric:=(

λAC = 400 × 10^-6;
λE = 10 × 10^-6;
λA = 2 × 10^-6;

cAColc = 95/100;
cEolc = 9/10;

cAC1 = covCal[4, 1, λAC, 30];
cAC2 = covCal[4, 2, λAC, 30];
cAC3 = cAColc;

cE1 = covCal[4, 1, λE, 30];
cE2 = covCal[4, 2, λE, 30];
cE3 = cEolc;

r[λ_, t_]:=e^(-λt);

);

setSymbolic:=Clear[λAC, λE, λA, cE, cAC1, cAC2, cAC3, cAColc, cEolc,
 cE1, cE2, cE3, r, t];
```

The probability of actuation system failure for the FLC case is

```
setNumeric;
times = {1, 2, 5, 10, 20};
Table[{t = times[[i]], N[1 - rSysFLC, 6]},
{i, 1, Length[times]}]//TableForm
setSymbolic;
```

1	3.52365×10^{-11}
2	1.70004×10^{-10}
5	2.09061×10^{-9}
10	1.58991×10^{-8}
20	1.25477×10^{-7}

For the PFC case,

```
rSysPFC =
 rSysFLC/.cAC1 → 1/.cAC2 → 1/.cAC3 → 1/.cE1 → 1/.cE2 →
1/.cE3 → 1;
```

The first and last terms and the length of the resulting PFC reliability polynomial, `rSysPFC`, are

```
rSysPFC//Length
```
17

```
rSysPFC//First
```
$8\, r[\lambda A, t] r[\lambda AC, t] r[\lambda E, t]$

```
rSysPFC//Last
```
$-r[\lambda A, t]^2 r[\lambda AC, t]^4 r[\lambda E, t]^8$

```
setNumeric;
times = {1,2,5,10,20};
Table[{t = times[[i]], N[1 - rSysPFC, 6]},
{i, 1, Length[times]}]//TableForm
setSymbolic;
```

1	4.02557×10^{-12}
2	1.64089×10^{-11}
5	1.15935×10^{-10}
10	6.53956×10^{-10}
20	5.63108×10^{-9}

For the OLC case,

```
rSysOLC =
 rSysFLC/.cAC1 → 1/.cAC2 → 1/.cAC3 → cAColc/.cE1 → 1/.cE2 →
1/.cE3 → cEolc;
```

```
rSysOLC//Length
```
51

```
rSysOLC//First
```
$8\, \text{cAColc}\, \text{cEolc}\, r[\lambda A, t] r[\lambda AC, t] r[\lambda E, t]$

```
rSysOLC//Last
```
$-16\, \text{cEolc}^2\, r[\lambda A, t]^2 r[\lambda AC, t]^4 r[\lambda E, t]^8$

```
setNumeric;
times = {1,2,5,10,20};
Table[{t = times[[i]], N[1 - rSysOLC, 6]},
{i, 1, Length[times]}]//TableForm
setSymbolic;
```

1	1.92429×10^{-11}
2	1.38030×10^{-10}
5	2.01077×10^{-9}
10	1.57397×10^{-8}
20	1.25159×10^{-7}

5.4.2 Actuation Subsystem Analysis

Numerical values for the probability of failure of the actuation subsystem are shown in Table 5.5. These results were computed using symbolic expressions for PFC, FLC and OLC versions of the actuation subsystem and are shown graphically as a log-log plot in Figure 5.12.

Table 5.5 Actuation subsystem probability of failure

t (hrs)	PFC	FLC	OLC
1	4.02557×10^{-12}	3.52365×10^{-11}	1.92429×10^{-11}
2	1.64089×10^{-11}	1.70004×10^{-10}	1.38030×10^{-10}
5	1.15935×10^{-10}	2.09061×10^{-9}	2.01077×10^{-9}
10	6.53956×10^{-10}	1.58991×10^{-8}	1.57397×10^{-8}
20	5.63108×10^{-9}	1.25477×10^{-7}	1.25159×10^{-7}

The actuation subsystem is shown to be a reasonable match for the computer control subsystem analyzed in the previous section; the standalone reliability of the actuation subsystem is at least an order of magnitude better than that of the computer control subsystem. Note that for low mission times, however, the PFC actuation subsystem is less reliable than the computer control subsystem.

For the actuation subsystem, the FLC and OLC models are nearly coincident for mission times greater than 3 hours. The PFC model has a slope of nearly 4 dpd at high mission times and a slope approaching 2 dpd at low mission times; this behavior is the result of the lack of the actuation system's full quad redundancy. The actuators are only dual redundant, and even though they have a rather low failure rate of $\lambda = 2$ fpmh, this reduced level of redundancy starts to dominate for mission times less than about 5 hours. Also notice that the IFC effects reduce the system reliability by a factor of approximately five, again demonstrating the necessity of appropriate modeling of coverage effects.

5.5 Combined Computer and Actuation Systems

In principle, the combination of the computer control and actuation subsystem codes into a single model to determine the reliability of the full system described in Figures 5.10 and 5.11 is a straightforward task. The standalone computer control system polynomial (prior to making the i.i.d. substitution that transforms the expression into a function of $\boldsymbol{r}[\lambda, t]$) has 912 terms, and the actuation subsystem polynomial has 2,912 terms. Since BSV must operate on the basic reliability polynomials (prior to conversion to the much smaller $\boldsymbol{r}[\lambda, t]$ polynomials), the resulting polynomial for the combined computer/actuation system will be very large. Although *Mathematica* can handle rather large expressions, a polynomial with this many terms is too

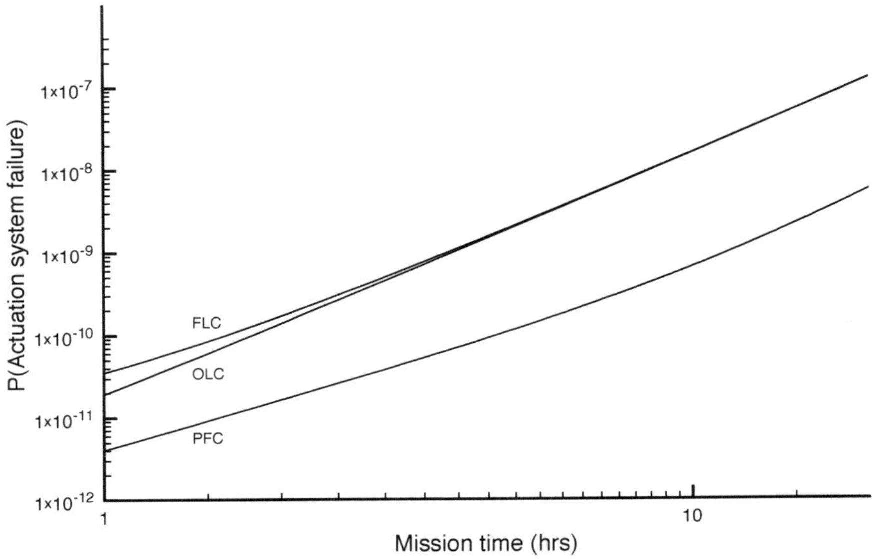

Fig. 5.12 Actuation subsystem probability of failure

large to be practical. Thus, to develop exact expressions for systems with more than 20–25 individual components, a different modeling approach is necessary.

The next chapter demonstrates the use of the *Mathematica* BSV tools in developing a conditional probability model (CPM) that is capable of modeling the entire system. Using CPM, it is possible to model systems of arbitrarily large size. Subsequent chapters also show how large systems with far more variables can be handled with a single integrated model using binary decision diagram (BDD) techniques. The BDD problem definition is very similar to the BSV formulation but is not limited to the range of 20–25 individual components.

References

1. Myers AF (2007) k-out-of-n:G System Reliability With Imperfect Fault Coverage. IEEE Trans Relia 56:464–473
2. Chang YR, Amari SV, Kuo SY (2002) Reliability Evaluation of Multi-state Systems Subject to Imperfect Coverage using OBDD. Proc Pac Rim Intl Symp Dependable Comp, IEEE
3. Doyle SA, Dugan LB, Patterson–Hine FA (1995) A combinatorial approach to modeling imperfect coverage. IEEE Trans Relia 44: 87–94
4. Amari SV, Dugan JB, Misra RB (1999) A separable method for incorporating imperfect fault-coverage in combinatorial models. IEEE Trans Relia 48: 267–274
5. Amari SV, Dugan JB, Misra RB (1999) Optimal reliability of systems subject to imperfect fault coverage. IEEE Trans Relia 48:275–284
6. Dugan JB (1989) Fault tree and imperfect coverage. IEEE Trans Relia 37:177–185
7. Dugan JB and Trivedi KS (1989) Coverage modeling for dependability analysis of tolerant systems. IEEE Trans Relia 37:775–787
8. Trivedi KS (2002) Probability and Statistics with Reliability, Queuing and Computer Science Applications, 2nd edn. Wiley, New York

Chapter 6
Conditional Probability Modeling Using BSV

Where there's a will, there's a way.

Abstract It was shown in Chapter 5 that the use of the BSV technique offers a general means of assessing complex system reliability. BSV, however, requires that the resulting polynomials be fully expanded, and for large systems, the resulting expressions can become too large to be practical. BSV techniques can, however, be used in the development of conditional probability models for the assessment of systems of arbitrary size and complexity, as long as the system can be represented in terms of disjoint expressions. This chapter demonstrates the use of BSV in the development of these conditional probability models.

6.1 Background

The reliability of a system can be characterized as a sum of disjoint products (SDP). This chapter illustrates the development of conditional probability models (CPM) for a combined system through the use of smaller subsystem models that allow the overall system reliability to be expressed as the sum of the products of the subsystem models. Assume some system (A) is made up two subsystems (B and C). Let rA be the reliability of A, rB be the reliability of B and rC be the reliability of C. If there exists, for the overall system, a set of n *mutually exclusive states*, then

$$rA = \sum_{i=1}^{n} rA_i \,. \tag{6.1}$$

If Subsystems B and C have no components in common, then

$$rA_i = rB_i \cdot rC_i \,, \tag{6.2}$$

where rB_i and rC_i are disjoint products. Also,

$$rA = \sum_{i=1}^{n} rB_i \cdot rC_i \,, \tag{6.3}$$

in which case the reliability of System A has been expressed as an SDP of its subsystem reliabilities.

To use an SDP strategy, it must be possible to identify a set of n mutually exclusive states of the system and its subsystems. Of course, if Subsystems B and C have neither common components *nor* common dependencies, then $n = 1$ is possible and $rA = rB \cdot rC$; that is, the system reliability is simply the product of the reliabilities of its subsystems.

6.2 Combined Computer and Actuation System CPM

The system depicted in Figures 5.10 and 5.11 in the preceding chapter can be modeled in a manner that supports an SDP. A naive approach to this model simply involves calculating the product of the expressions that represent the computer control and actuation subsystems, with the condition that they have no common components. This approach, though appealing in its simplicity, does not yield correct results. Despite having no common components, the two subsystem models are not disjoint. Both the computer control and actuation subsystems have a common dependence on the electrical bus output, and the actuation subsystem is dependent on the number of operational channels in the computer control subsystem. Both the computer control and actuation subsystem components are powered by the same set of buses, and the actuation computers receive command data only from their local control computer. This common dependence means that the two subsystems are not disjoint.

Either of these two interdependencies (the number of operational buses or the number of operational channels) can, however, be exploited in the development of a CPM, using SDP to provide a full-system model. Models for both the computer control and actuation subsystems can be developed for the following mutually exclusive conditions: the exact number of operational power buses, i, where $i \in \{1, \ldots, 4\}$, or the exact number of operational channels, j, where $j \in \{1, \ldots, i\}$ (the number of operational channels is equal to the number of operational computers in the computer subsystem). Thus, if rCB_i is the computer system reliability given that exactly i of the $nBus = 4$ buses are operational, then the reliability of the computer control subsystem is

$$rCompSys = \sum_{i=1}^{nBus} rCB_i \, . \tag{6.4}$$

By the same rationale, given that rCC_j is the computer system reliability and that exactly j channels are operational,

$$rCompSys = \sum_{j=1}^{nChn} rCC_j \, . \tag{6.5}$$

Note that $rCB_i \neq rCC_i$, since it is possible that the number of operational buses exceeds the number of operational channels. Although their sums are equal, the individual rCB and rCC terms are different.

In a similar fashion, if rAB_i is the actuation subsystem reliability given i operational buses and if rAC_j is the actuation subsystem reliability given j operational channels, then

$$rActSys = \sum_{i=1}^{nBus} rAB_i \tag{6.6}$$

and

$$rActSys = \sum_{j=1}^{nChn} rAC_j \, . \tag{6.7}$$

The overall system reliability, including both the computer control subsystem and the actuation subsystem, is the sum of the following disjoint products:

$$rSys = \sum_{i=1}^{nBus} rCB_i \cdot rAB_i \tag{6.8}$$

and

$$rSys = \sum_{j=1}^{nChn} rCC_j \cdot rAC_j \, . \tag{6.9}$$

The required subsystem reliability expressions (which, in this case, are conditioned on either the number of operational channels or the number of operational buses) can be derived using BSV in a fashion that starts with subsystem reliability expressions that are obtained in the same way as those presented in Chapter 5. Overall system reliability can be determined by the following method:

1. Derive expressions for the reliability of each subsystem using BSV or some other appropriate technique.
2. Derive the $1, \ldots, n$ reliability expressions for each subsystem for exactly the mutually exclusive (disjoint) conditions.
3. Calculate the product of the subsystem expressions, each summed over the n conditions, to get the overall system reliability.

If the conditional is directly observable within the subsystem to be modeled, then Step 2 can be readily implemented using the **ReX** functions presented in Chapters 3 and 4 for determining the probability that *exactly* k out of n elements are operational. For instance, the number of operational buses in the computer subsystem discussed above is directly observable in the computer subsystem. Once the computer subsystem reliability is derived, the probability that exactly $1, \ldots, nBus$ buses are operational can be computed using the **ReX** function. These expressions can then be used as the rCB_i terms in Equation (6.8). The following code is a modification of the computer subsystem code developed in Section 5.3.1.

```
sP = {P1,P2,P3,P4};
sS = {S1,S2,S3,S4};
sC = {C1,C2,C3,C4};

cS = {cS1,cS2,cS3};
cC = {cC1,cC2,cC3};

sBout = {P1 ⊕ P4,P2 ⊕ P1,P3 ⊕ P2,P4 ⊕ P3};

sSpwr = sS ⊗ sBout;
sCpwr = sC ⊗ sBout;

sSvoted = RaL[1,sSpwr ⊗ sC,cS];
sCvoted = RaL[1,sCpwr,cC];
```

Calculate the computer subsystem reliability:

```
sSysComp = sSvoted ⊗ sCvoted;
```

Develop expressions for exactly $1,\ldots,nBus$ operational buses:

```
sSysComp1 = sSysComp ⊗ ReX[1,sBout];
sSysComp2 = sSysComp ⊗ ReX[2,sBout];
sSysComp3 = sSysComp ⊗ ReX[3,sBout];
sSysComp4 = sSysComp ⊗ ReX[4,sBout];
```

In the case of the actuation subsystem, however, the state of the bus is not directly observable (the bus state is modeled as a portion of the computer subsystem), and the derivation of the conditional expressions for the actuation subsystem requires an additional step. Modeling the effect of the actuation system components' dependence on the bus can be accomplished using a proxy variable for the bus state. Once the derivation is complete, the proxy variables are removed using substitution.

The actuation subsystem code from Section 5.4.1 has been modified, as shown below, to include the bus proxy vector **sB = {B1,B2,B3,B4}**, thus allowing derivation of the FLC expressions for the actuation system reliability given exactly $1,\ldots,nBus$ operational buses.

```
sB = {B1,B2,B3,B4};

sAC = {AC1,AC2,AC3,AC4};
sEL = {EL1,EL2,EL3,EL4};
sER = {ER1,ER2,ER3,ER4};
```

```
cAC = {cAC1, cAC2, cAC3};
cEL = {cEL1, cEL2, cEL3};
cER = {cER1, cER2, cER3};

sACvoted = RaL[1, sAC ⊗ sB, cAC];
sELselect = sACvoted ⊗ RaL[1, sEL ⊗ sAC ⊗ sB, cEL];
sERselect = sACvoted ⊗ RaL[1, sER ⊗ sAC ⊗ sB, cER];
sACT1out = A1 ⊗ sELselect;
sACT2out = A2 ⊗ sERselect;
sSysACT = sACT1out ⊕ sACT2out;
```

Using substitution, the proxy variables are removed from the resulting conditional expressions for the probability that exactly 1,...,*nBus* buses are operational:

```
sSysACT1 = sSysACT/.B1 → 1/.B2 → 0/.B3 → 0/.B4 → 0;
sSysACT2 = sSysACT/.B1 → 1/.B2 → 1/.B3 → 0/.B4 → 0;
sSysACT3 = sSysACT/.B1 → 1/.B2 → 1/.B3 → 1/.B4 → 0;
sSysACT4 = sSysACT/.B1 → 1/.B2 → 1/.B3 → 1/.B4 → 1;
```

Here, **sSysACT4** is the reliability of the actuation system given that four buses are operational, **sSysACT3** is the reliability of the actuation system given that three buses are operational and so on.

Using the derivation approach shown above, the reliability of the overall system, which is a combination of the computer and actuation subsystems, can be computed using Equation (6.8). Note that all of the calculations for deriving the conditional expressions must be performed prior to applying the i.i.d. transformation and prior to converting the results to $r[\lambda, t]$ form. Once the conditional expressions have been derived, however, the expressions used in Equation (6.8) can (and should) be converted to $r[\lambda, t]$ form, which substantially reduces the polynomial size. For this example, the conversion is accomplished using the **TOr** substitution rule:

```
TOr = {
P1 → r[λP, t], P2 → r[λP, t], P3 → r[λP, t], P4 → r[λP, t],
S1 → r[λS, t], S2 → r[λS, t], S3 → r[λS, t], S4 → r[λS, t],
C1 → r[λC, t], C2 → r[λC, t], C3 → r[λC, t], C4 → r[λC, t],
AC1 → r[λAC, t], AC2 → r[λAC, t], AC3 → r[λAC, t], AC4 → r[λAC, t],
EL1 → r[λE, t], EL2 → r[λE, t], EL3 → r[λE, t], EL4 → r[λE, t],
ER1 → r[λE, t], ER2 → r[λE, t], ER3 → r[λE, t], ER4 → r[λE, t],
cEL1 → cE1, cEL2 → cE2, cEL3 → cE3,
cER1 → cE1, cER2 → cE2, cER3 → cE3,
A1 → r[λA, t], A2 → r[λA, t]};
```

Using the **TOr** substitution rule, the conditional expressions for the computer and actuation subsystems can be converted to $r[\lambda, t]$ form:

```
rSysComp1 = sSysComp1/.TOr;
rSysComp2 = sSysComp2/.TOr;
rSysComp3 = sSysComp3/.TOr;
rSysComp4 = sSysComp4/.TOr;

rSysACT1 = sSysACT1/.TOr;
rSysACT2 = sSysACT2/.TOr;
rSysACT3 = sSysACT3/.TOr;
rSysACT4 = sSysACT4/.TOr;
```

The overall FLC system reliability consisting of both the computer subsystem and a single actuation subsystem is calculated using the equivalent of Equation (6.8):

```
rSys1Surf =
((rSysComp1 × rSysACT1) + (rSysComp2 × rSysACT2) +
(rSysComp3 × rSysACT3) + (rSysComp4 × rSysACT4))//Expand;
```

The resulting fully expanded polynomial for the combined system has 10,095 terms.

```
rSys1Surf//Length
10095
```

A check of the first and last terms of the resulting polynomial shows the expected form; the first term represents a minimum path of the system's components, and the last term includes the full complement of components (see Section 5.3.1 for a full discussion).

rSys1Surf//First

32 cAC1 cAC2 cAC3 cC1 cC2 cC3 cE1 cE2 cE3 cS1 cS2 cS3 $r[\lambda\text{A}, t] r[\lambda\text{AC}, t] r[\lambda\text{C}, t] r[\lambda\text{E}, t] r[\lambda\text{P}, t] r[\lambda\text{S}, t]$

rSys1Surf//Last

$-64\ \text{cE1}^2\ \text{cE2}^2\ \text{cE3}^2\ \text{cS1 cS2 cS3}\ r[\lambda\text{A}, t]^2 r[\lambda\text{AC}, t]^4 r[\lambda\text{C}, t]^4 r[\lambda\text{E}, t]^8 r[\lambda\text{P}, t]^4 r[\lambda\text{S}, t]^4$

This section has illustrated a CPM that was developed for the reliability of a system having too many components to be derived using only BSV techniques. Note that all of the calculations for deriving the conditional expressions are performed prior to applying the i.i.d. transformation and prior to converting the results to **r[λ, t]** format (both done using the **TOr** substitution rule).

Although this example involved an FLC model of the combined computer and actuation system, the derivation of PFC and OLC expressions can be done in an analogous fashion by changing (or eliminating, in the PFC case) the coverage arguments used in the **RaL** functions.

6.3 Combined System CPM Results

An expression for the reliability of an FLC version of the combined computer and actuation system was developed in the previous section. This expression, along with corresponding expressions for PFC and OLC versions of the system, was used to compute the probability of system failure values given in Table 6.1. These results are also shown as a log-log plot in Figure 6.1.

Table 6.1 Combined computer and actuation system probability of failure

t (hrs)	PFC	FLC	OLC
1	4.20552×10^{-12}	1.88866×10^{-10}	1.40635×10^{-10}
2	1.92855×10^{-11}	1.20577×10^{-9}	1.10934×10^{-9}
5	2.28002×10^{-10}	1.74365×10^{-8}	1.71957×10^{-8}
10	2.43906×10^{-9}	1.37806×10^{-7}	1.37325×10^{-7}
20	3.39384×10^{-8}	1.10039×10^{-6}	1.09943×10^{-6}

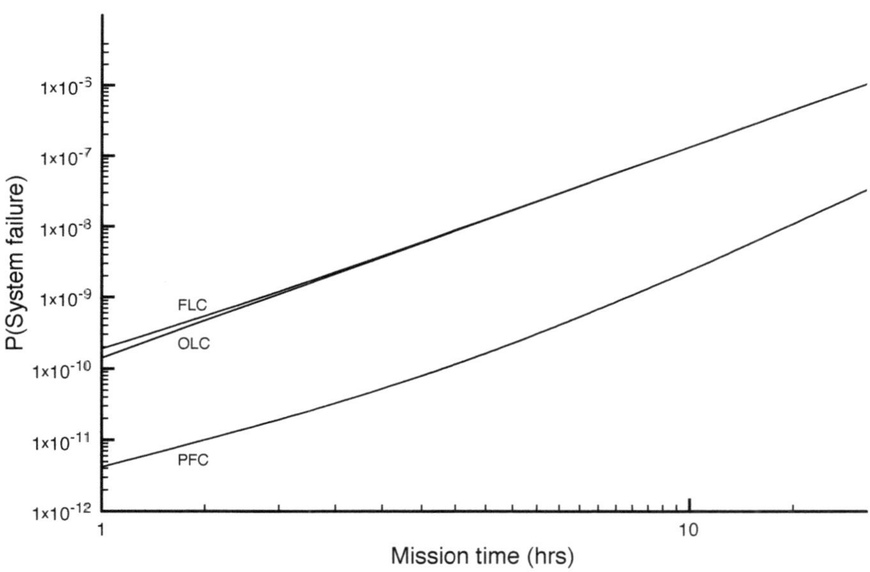

Fig. 6.1 Computer and single actuation system probability of failure

Again, the significant effect of IFC is apparent in these results. For mission times less than about 4 hours, the FLC system has a probability of failure that is at least an order of magnitude greater than that of the PFC system. The effect of the dual redundancy of the actuators is still apparent, as seen in the reduction of slope for the PFC system at low mission times. This effect, however, is less obvious than it is in the plot of the actuation system alone, which is shown in Figure 5.12. For mission times greater than 5 hours, OLC provides an excellent approximation of the FLC system. The substantially similar character of the combined system plot in comparison with that of the standalone computer subsystem plot shown in Figure 5.9 is an indication that the actuation subsystem is a reasonable match for the computer control system.

The next section demonstrates the use of the same techniques in the derivation of a CPM for a combined system with multiple actuation systems. The full *Mathematica* BSV code for these multiple actuation system models is presented in the next section; a subset of this code reproduces the results given in the present section.

6.4 CPM for a System with Multiple Actuation Subsystems: System A

The system described in Figures 5.10 and 5.11 and whose probability of failure is shown in Figure 6.1 might well represent a simple—albeit highly reliable—computerized feedback motion control system for a manufacturing plant. It might also represent a digital system for the single-axis (in fact, single-surface) control of an aircraft. Since it is highly unlikely that such a system would be developed for a single aircraft control surface, it is of interest to expand the model to handle multiple control surfaces.

For an aircraft with six control surfaces (all of which are required for flight), each surface has an associated actuation computer and is powered by two actuators, as shown in Figure 6.2. The computers AC_i associated with each control surface use their CCDL to implement an input voting plane. As a result, each AC_i contributes to system reliability even if its local C_i has failed.[1] Additionally, both of the actuators associated with each control surface have a quad set of servo-electronics. This aircraft control system can be modeled by way of a simple modification to Equation (6.8). Since each of the six control surfaces consists of distinct and disjoint components, the only modification required to obtain the reliability of the new system is to raise the rAB_i term to the sixth power:

$$rSys_A = \sum_{i=1}^{nBus} rCB_i \cdot (rAB_i)^6 \ . \tag{6.10}$$

This difference is a consequence of having six actuation subsystems (surfaces) instead of a single actuation subsystem. The results already derived in Section 6.2 can be used to generate a model for the system with six control surfaces:

---

[1] An arbitrary member of a redundant set of components is referred to as X_i; for example, an arbitrary computer is labeled C_i.

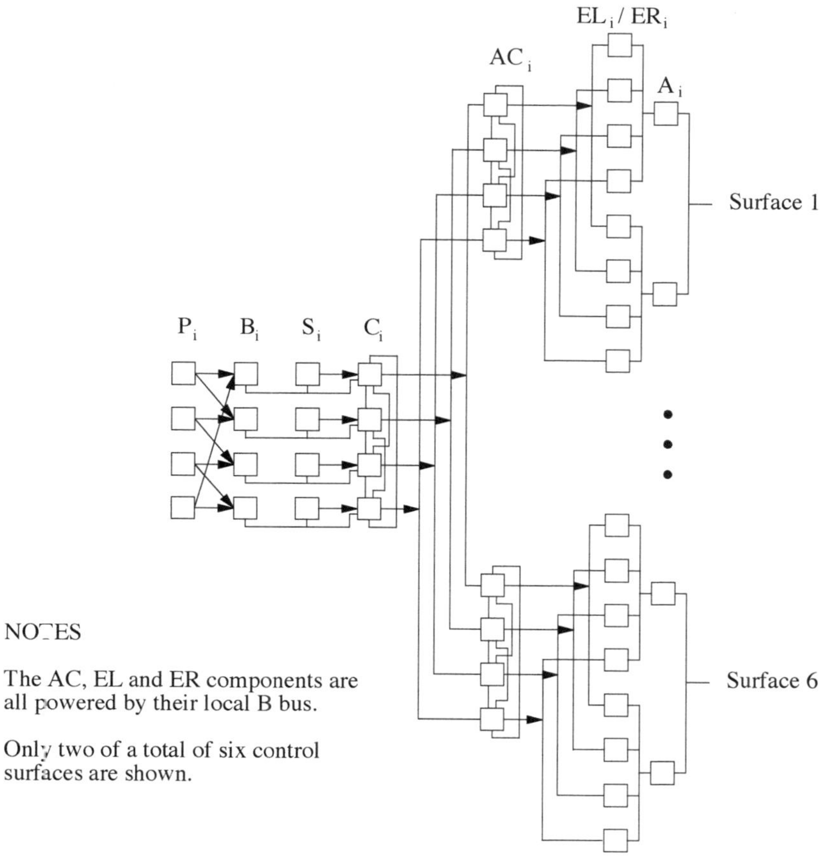

Fig. 6.2 Quad digital flight control system with six control surfaces—System A

rSys6Surf =
((**rSysComp1** × **rSysACT1**^{6}) + (**rSysComp2** × **rSysACT2**^{6}) +
(**rSysComp3** × **rSysACT3**^{6}) + (**rSysComp4** × **rSysACT4**^{6}));

The only difference between this expression and the previous expression for the system with a single actuation subsystem is, of course, in the **rSysACTn** terms, which are raised to the sixth power. In this section, however, a somewhat more compact derivation is illustrated. The results are nevertheless equivalent.

The *Mathematica* BSV code given below derives the *rCB* and *rAB* expressions for Equation (6.10). This code implements a CPM of the system shown in Figure 6.2, with the subsystem models being conditional on the number of operating buses; that is, the CPM for the computer subsystem (the P, S and C components) are derived for 1,...,*nBus* operational buses, and the same is done for the control surface subsystem (the AC and EL/ER components). The code derives expressions for an FLC version of the system and then derives PFC and OLC models from the FLC expression.

The function **GenRSet** generates a list of redundant component variables:

```
GenRSet[varName_, lng_]:=Module[{i},
Table[ToExpression[ToString[varName] <> ToString[i]], {i, lng}]];
```

For example,

```
GenRSet[A, 4]
{A1, A2, A3, A4}
```

The numerical data for the systems analyzed in this chapter are the same as those used in Chapter 5 (Tables 5.1, 5.2 and 5.4); these data are repeated in the **setNumeric** routine below. Once a reliability model has been created for a system it is useful to be able to perform both symbolic and numeric evaluations. The following are routines for conversion between numerical and symbolic computations:

```
setNumeric:=(

λP = 200 × 10^-6;
λS = 250 × 10^-6;
λC = 400 × 10^-6;
λAC = 400 × 10^-6;
λE = 10 × 10^-6;
λA = 2 × 10^-6;

cSolc = 9/10;
cColc = 95/100;
cACoIc = 95/100;
cEolc = 9/10;

cS1 = covCal[4, 1, λS, 30]; cS2 = covCal[4, 2, λS, 30];
cS3 = cSolc;

cC1 = covCal[4, 1, λC, 30]; cC2 = covCal[4, 2, λC, 30];
cC3 = cColc;

cAC1 = covCal[4, 1, λAC, 30]; cAC2 = covCal[4, 2, λAC, 30];
cAC3 = cAColc;

cE1 = covCal[4, 1, λE, 30]; cE2 = covCal[4, 2, λE, 30];
cE3 = cEolc;
```

```
r[λ_, t_]:=e^(-λt);

);

setSymbolic:=Clear[λP, λS, λC, λAC, λE, λA, cS, cC, cAC, cE, cEL, cER,
cSolc, cColc, cAColc, cEolc, cS1, cS2, cS3, cC1, cC2, cC3, cAC1, cAC2,
cAC3, cE1, cE2, cE3, r, t];
```

The following are substitution rules for conversion to polynomial form in `r[λ, t]`:

```
TOr = {
P1 → r[λP, t], P2 → r[λP, t], P3 → r[λP, t], P4 → r[λP, t],
S1 → r[λS, t], S2 → r[λS, t], S3 → r[λS, t], S4 → r[λS, t],
C1 → r[λC, t], C2 → r[λC, t], C3 → r[λC, t], C4 → r[λC, t],
AC1 → r[λAC, t], AC2 → r[λAC, t],
AC3 → r[λAC, t], AC4 → r[λAC, t],
E11 → r[λE, t], E12 → r[λE, t], E13 → r[λE, t], E14 → r[λE, t],
E21 → r[λE, t], E22 → r[λE, t], E23 → r[λE, t], E24 → r[λE, t],
cE11 → cE1, cE12 → cE2, cE13 → cE3, cE14 → cE4,
cE21 → cE1, cE22 → cE2, cE23 → cE3, cE24 → cE4,
A1 → r[λA, t], A2 → r[λA, t]};
```

Assure symbolic calculations with the following statement:

```
setSymbolic;
```

The computer subsystem variables are

```
nChn = 4;
nBus = 4;

sP = GenRSet[P, nChn];
sS = GenRSet[S, nChn];
cS = GenRSet[cS, nChn - 1];
sC = GenRSet[C, nChn];
cC = GenRSet[cC, nChn - 1];
```

The computer subsystem model is

```
sBout = {P1 ⊕ P4, P2 ⊕ P1, P3 ⊕ P2, P4 ⊕ P3};
sSpwr = sS ⊗ sBout;
sCpwr = sC ⊗ sBout;
sSin = sSpwr ⊗ sC;
```

```
sSvoted = RaL[1, sSin, cS];
sCvoted = sSvoted ⊗ RaL[1, sCpwr, cC];
```

The term **sCvoted** is the reliability of the computer subsystem.

Next, create a table for the computer subsystem reliability with 1,..., *nBus* operational buses. This table corresponds to the rCB_i terms in Equation (6.8):

```
rThruC = Table[0, {nBus}];

For[iBus = 1, iBus ≤ nBus, iBus++,
rThruC[[iBus]] = sCvoted ⊗ ReX[iBus, sBout]/.TOr;
];
```

The term **rThruCsum** should also be equal to the computer subsystem reliability.

```
rThruCsum = Σ_{i=1}^{nBus} rThruC[[i]];
```

Check to make sure that these expressions are equal:

```
rThruCsum==(sCvoted/.TOr)
True

rThruCsum//Length
78

rThruCsum//First
8 cC1 cC2 cC3 cS1 cS2 cS3 r[λC, t] r[λP, t] r[λS, t]

rThruCsum//Last
4 cS1 cS2 cS3 r[λC, t]^4 r[λP, t]^4 r[λS, t]^4
```

The accurate (to nine significant digits) calculation of FLC reliability requires extended-precision calculation of the FLC coverage values—25 digits in this case.

```
covCal[n_, m_, λ_, w_]:=
Module[{wHr = w/(1000 × 3600)},
N[e^(-(n-m)λwHr), 25]
];
```

Calculate the computer subsystem reliability expression:

```
SysC1 = Σ_{i=1}^{nBus} rThruC[[i]];
```

Generate a table of numerical values for the computer subsystem probability of failure:

```
setNumeric;
times = {1,2,5,10,20};
Table[{t = times[[i]], N[1 - SysC1,6]}, {i,1,Length[times]}]//
TableForm
setSymbolic;
```

1	1.53622×10^{-10}
2	1.03566×10^{-9}
5	1.53420×10^{-8}
10	1.21845×10^{-7}
20	9.73925×10^{-7}

For the actuation subsystem model, expressions for 1,. . . , *nBus* operational buses are needed. From the perspective of the actuation subsystem, the number of operational buses is equivalent to the number of operational actuation subsystem channels, since the servo-electronics components are each powered by their local bus.

The approach used below involves the derivation of expressions for the reliability of the actuation system for **iBus** = 1,. . . , *nBus* using proxy variables B1,. . .,B4, which have an effect equivalent to that of the local bus. The proxy variables are eliminated from the expression using the substitution rule **BusPat,** which, for **iBus** = 4, has the value

{B1→1, B2→1, B3→1, B4→1}

and for **iBus** = 3,

{B1→1, B2→1, B3→1, B4→0}

and so on.

In the following code, the set of utility routines defined below is used. The **GenSubRule** routine automates the generation of a set of substitution rules; this routine is used in conjunction with **GenPat** to generate the required substitution rules to eliminate the **B** proxy variables as a function of the number of operational channels.

```
GenSubRule[varName_,pat_:List]:=
Module[{n = Length[pat]},

Table[Rule[ToExpression[ToString[varName] <> ToString[i]],
pat[[i]]], {i, n}]
];

GenPat[n_,k_]:=Module[{},
result = Table[0, {n}];
```

```
For[i = 1, i ≤ k, i++,
result[[i]] = 1;
];
result
];
```

For example,

```
GenSubRule[B, GenPat[4, 3]]
{B1 → 1, B2 → 1, B3 → 1, B4 → 0}
```

Derive a table of actuation subsystem expressions for exactly 1,..., *nBus* operational buses:

```
setSymbolic;
sAC = GenRSet[AC, nChn];
sEL = GenRSet[E1, nChn];
sER = GenRSet[E2, nChn];

cAC = GenRSet[cAC, nChn − 1];
cEL = GenRSet[cE1, nChn − 1];
cER = GenRSet[cE2, nChn − 1];

sB = GenRSet[B, nChn];

sACin = sAC ⊗ sB;
sACvoted:=RaL[1, sACin, cAC];
sELselect = sACvoted ⊗ RaL[1, sEL ⊗ sACin, cEL];
sERselect = sACvoted ⊗ RaL[1, sER ⊗ sACin, cER];
sACT1out = A1 ⊗ sELselect;
sACT2out = A2 ⊗ sERselect;
sActSys = sACT1out ⊕ sACT2out;

rActSys = Table[0, {nChn}];
For[jChn = 1, jChn ≤ nChn, jChn++,
BusPat = GenSubRule[B, GenPat[nChn, jChn]];

rActSys[[jChn]] = sActSys/.BusPat/.TOr;
];
sysA1 = (∑_{i=1}^{nBus} rThruC[[i]]rActSys[[i]]) //Expand;

sysA1//Length
10095
```

The first and last terms of the resulting **sysA1** expression are, of course, identical to those obtained in Section 6.2 using the alternative derivation.

```
sysA1//First
```
32 cAC1 cAC2 cAC3 cC1 cC2 cC3 cE1 cE2 cE3 cS1 cS2 cS3
$r[\lambda\text{A}, t]\ r[\lambda\text{AC}, t]\ r[\lambda\text{C}, t]\ r[\lambda\text{E}, t]\ r[\lambda\text{P}, t]\ r[\lambda\text{S}, t]$

```
sysA1//Last
```
$-64\ \text{cE1}^2\ \text{cE2}^2\ \text{cE3}^2\ \text{cS1 cS2 cS3}\ r[\lambda\text{A}, t]^2$
$r[\lambda\text{AC}, t]^4\ r[\lambda\text{C}, t]^4\ r[\lambda\text{E}, t]^8\ r[\lambda\text{P}, t]^4\ r[\lambda\text{S}, t]^4$

It is more numerically efficient to create a polynomial that has not been expanded; the following expression implements Equation (6.8) in such a manner:

sysA1 = $\sum_{i=1}^{\text{nBus}}$ **rThruC[[*i*]] rActSys[[*i*]];**

Generate a table of numerical values for the combined FLC system reliability:

```
setNumeric;
times = {1,2,5,10,20};
Table[{t = times[[i]], N[1 - sysA1,6]}, {i,1,Length[times]}]//
TableForm
setSymbolic;
```

1	1.88866×10^{-10}
2	1.20577×10^{-9}
5	1.74365×10^{-8}
10	1.37806×10^{-7}
20	1.10039×10^{-6}

PFC and OLC models can be derived from the FLC result by making the appropriate substitutions for the coverage variables:

```
TOpfc = {
cS1 → 1, cS2 → 1, cS3 → 1,
cC1 → 1, cC2 → 1, cC3 → 1,
cAC1 → 1, cAC2 → 1, cAC3 → 1,
cE1 → 1, cE2 → 1, cE3 → 1
};

TOolc = {
cS1 → 1, cS2 → 1, cS3 → cSolc,
cC1 → 1, cC2 → 1, cC3 → cColc,
cAC1 → 1, cAC2 → 1, cAC3 → cACoolc,
cE1 → 1, cE2 → 1, cE3 → cEolc
};
```

Symbolic models for PFC and OLC systems can be derived using these substitution rules:

```
sysA1flc = sysA1;
sysA1pfc = sysA1/.TOpfc;
sysA1olc = sysA1/.TOolc;
sysA1pfcEx = sysA1pfc//Expand;

sysA1pfcEx//Length
181

sysA1pfcEx//First
32 r[λA, t] r[λAC, t] r[λC, t] r[λE, t] r[λP, t] r[λS, t]

sysA1pfcEx//Last
-r[λA, t]^2 r[λAC, t]^4 r[λC, t]^4 r[λE, t]^8 r[λP, t]^4 r[λS, t]^4

sysA1olcEx = sysA1olc//Expand;

sysA1olcEx//Length
1661

sysA1olcEx//First
32 cACol c cColc cEolc cSolc r[λA, t] r[λAC, t] r[λC, t] r[λE, t] r[λP, t] r[λS, t]

sysA1olcEx//Last
-64 cEolc^2 cSolc r[λA, t]^2 r[λAC, t]^4 r[λC, t]^4 r[λE, t]^8 r[λP, t]^4 r[λS, t]^4
```

The PFC and OLC models could also have been derived in a direct fashion similar to that used for the FCL model.

Numerical results for PFC, FLC and OLC single-surface models (for the actuation subsystems) are computed as follows:

```
setNumeric;
times = {1, 2, 5, 10, 20};
Table[{t = times[[i]], N[1 - sysA1pfc, 6], N[1 - sysA1flc, 6],
N[1 - sysA1olc, 6]}, {i, 1, Length[times]}]//TableForm
setSymbolic;
```

1	4.20552×10^{-12}	1.88866×10^{-10}	1.40635×10^{-10}
2	1.92855×10^{-11}	1.20577×10^{-9}	1.10934×10^{-9}
5	2.28002×10^{-10}	1.74365×10^{-8}	1.71957×10^{-8}
10	2.43906×10^{-9}	1.37806×10^{-7}	1.37325×10^{-7}
20	3.39384×10^{-8}	1.10039×10^{-6}	1.09943×10^{-6}

To compute the reliability for a system with six surfaces, use the following expression, which implements Equation (6.10):

```
sysA6 = ∑_{i=1}^{nBus} rThruC[[i]]rActSys[[i]]^6;
```

```
setNumeric;
times = {1,2,5,10,20};
Table[{t = times[[i]], N[1 − sysA6,6]}, {i,1,Length[times]}]//
TableForm
setSymbolic;
```

1	3.65087×10^{-10}
2	2.05631×10^{-9}
5	2.79089×10^{-8}
10	2.17609×10^{-7}
20	1.73272×10^{-6}

The following are symbolic models for the PFC and OLC versions of the System A configuration with six surfaces:

```
sysA6flc = sysA6;
sysA6pfc = sysA6/.TOpfc;
sysA6olc = sysA6/.TOolc;
```

The numerical results for the PFC, FLC and OLC models are

```
setNumeric;
times = {1,2,5,10,20};
Table[{t = times[[i]], N[1 − sysA6pfc,6], N[1 − sysA6flc,6],
N[1 − sysA6olc,6]}, {i,1,Length[times]}]//TableForm
setSymbolic;
```

1	2.43335×10^{-11}	3.65087×10^{-10}	2.36880×10^{-10}
2	1.01332×10^{-10}	2.05631×10^{-9}	1.79998×10^{-10}
5	8.07917×10^{-10}	2.79089×10^{-8}	2.72687×10^{-8}
10	5.71642×10^{-9}	2.17609×10^{-7}	2.16331×10^{-7}
20	6.23329×10^{-8}	1.73272×10^{-6}	1.73017×10^{-6}

The symbolic results for the six-surface PFC model of System A are

```
sysA6pfcEx = sysA6pfc//Expand;
```

```
sysA6pfcEx//Length
```

12271

```
sysA6pfcEx//First
```

$32768\ r[\lambda A, t]^6\ r[\lambda AC, t]^6\ r[\lambda C, t]\ r[\lambda E, t]^6\ r[\lambda P, t]\ r[\lambda S, t]$

sysApfcEx//Last
$r[\lambda\text{A}, t]^{12}\ r[\lambda\text{AC}, t]^{24}\ r[\lambda\text{C}, t]^{4}\ r[\lambda\text{E}, t]^{48}\ r[\lambda\text{P}, t]^{4}\ r[\lambda\text{S}, t]^{4}$

The symbolic results for the System A OLC model are

sysAolcEx = sysAolc//Expand;

sysAolcEx//Length
1592748

sysAolcEx//First
$32768\ \text{cAColc}^{6}\ \text{cColc}\ \text{cEolc}^{6}\ \text{cSolc}\ r[\lambda\text{A}, t]^{6}$
$r[\lambda\text{AC}, t]^{6}\ r[\lambda\text{C}, t]\ r[\lambda\text{E}, t]^{6}\ r[\lambda\text{P}, t]\ r[\lambda\text{S}, t]$

sysAolcEx//Last
$67108864\ \text{cEolc}^{12}\ \text{cSolc}\ r[\lambda\text{A}, t]^{12}$
$r[\lambda\text{AC}, t]^{24}\ r[\lambda\text{C}, t]^{4}\ r[\lambda\text{E}, t]^{48}\ r[\lambda\text{P}, t]^{4}\ r[\lambda\text{S}, t]^{4}$

The CPM for the combined system with six control surfaces, which is modeled above, is a straightforward extension of the CPM developed in Section 6.2 using BSV for a combined system with a single actuation subsystem. One characteristic of CPM is the ease with which the number of occurrences of a given subsystem can be changed once the model is developed for the individual subsystems (computer and surface/actuation subsystems in this case). Once the single-surface subsystem CPM was derived, simply raising the subsystem expression to the sixth power changed the expression from a single-surface model to a six-surface model. It is shown in the next section, however, that a change in architectural arrangement may require a different conditional basis upon which the model is to be built.

6.5 CPM for a System with Multiple Actuation Subsystems: System B

A second (and perhaps more likely) configuration of the system with six control surfaces is shown in Figure 6.3. For this configuration, a single actuation computer provides commands to all six control surfaces. This system requires a CPM that treats the actuation computer elements, AC, separately from the servo-electronics and actuator models. Developing a CPM for this configuration requires a change in the conditional basis of the computer subsystem model. For System A, the number of operational channels in the surface/actuation subsystem is determined by the number of operational buses in the computer subsystem. For System B, however, the number of surface subsystem channels is determined by the number of operational AC_i components; this is because the EL/ER components communicate only with their local AC, and if the local AC has failed, the corresponding EL/ER components cannot contribute to overall system reliability. Table 6.2 summarizes the conditional basis for both the System A and System B subsystems.

Table 6.2 Conditional basis for Systems A and B

Subsystem	System A	System B
Computer	Number of operational buses	Number of operational AC_i
Surface	Number of effective channels	Number of effective channels

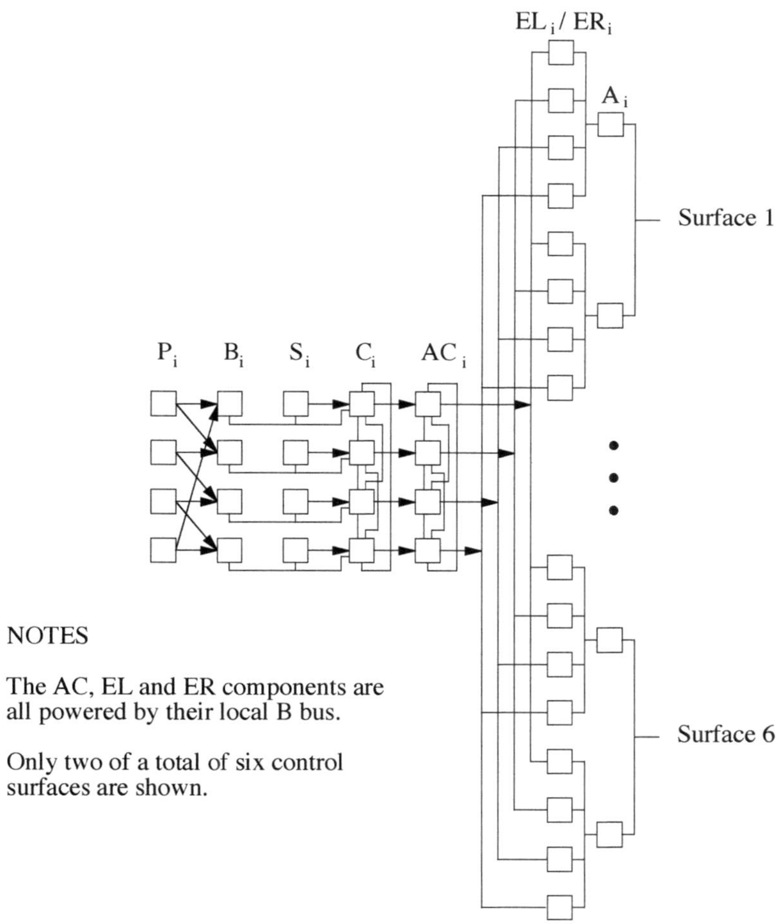

Fig. 6.3 Quad digital flight control system with six control surfaces—System B

If rCA_i represents computer subsystem reliability given that exactly i AC components are operational and that rAC_i is the expression for the effective number of operational channels in the surface subsystem, then the CPM for this system, $rSys_B$, is

$$rSys_B = \sum_{i=1}^{nChn} rCA_i \cdot (rAC_i)^6 \,. \tag{6.11}$$

The derivation of the System B subsystem expressions used in Equation (6.11) is done in a fashion similar to that used for the System A subsystems discussed in the previous section. The key difference between the two codes is that the models required for the computer subsystem and the control surface subsystems in System B are now conditioned on the number of operational channels, *nChn*, instead of the number of operational buses, *nBus*, which was used for System A.

An FLC version of System B based on Equation (6.11), is

```
setSymbolic;

nChn = 4;
nBus = 4;

sP = GenRSet[P, nChn];
sS = GenRSet[S, nChn];
sC = GenRSet[C, nChn];
sAC = GenRSet[AC, nChn];

cS = GenRSet[cS, nChn − 1];
cC = GenRSet[cC, nChn − 1];
cAC = GenRSet[cAC, nChn − 1];

sBout = {P1 ⊕ P4, P2 ⊕ P1, P3 ⊕ P2, P4 ⊕ P3};
sSpwr = sS ⊗ sBout;
sCpwr = sC ⊗ sBout;
sACpwr = sAC ⊗ sBout;
rThruAC = Table[0, {nChn}];

For[iChn = 1, iChn ≤ nChn, iChn++,
rThruAC[[iChn]] = sACvoted ⊗ ReX[iChn, sACpwr]/.TOr;
];

sEL = GenRSet[E1, nChn];
sER = GenRSet[E2, nChn];

cEL = GenRSet[cE1, nChn − 1];
cER = GenRSet[cE2, nChn − 1];
```

```
sB = GenRSet[B, nChn];

sELselect = RaL[1, sEL ⊗ sB, cEL];
sERselect = RaL[1, sER ⊗ sB, cER];
sACT1out = A1 ⊗ sELselect;
sACT2out = A2 ⊗ sERselect;
sActSys = sACT1out ⊕ sACT2out;

rActSys = Table[0, {nChn}];

For[jChn = 1, jChn ≤ nChn, jChn++,
BusPat = GenSubRule[B, GenPat[nChn, jChn]];
rActSys[[jChn]] = sActSys/.BusPat/.TOr;
];
```

sysB6 = $\sum_{i=1}^{\mathrm{nChn}}$ rThruAC[[i]]rActSys[[i]]^6;

Again, PFC and OLC models can be derived from these FLC results:

```
sysB6flc = sysB6;
sysB6pfc = sysB6/.TOpfc;
sysB6olc = sysB6/.TOolc;
```

Expand the expressions for the System B PFC, FLC and OLC results:

```
sysB6flcEx = sysB6flc//Expand;
sysB6pfcEx = sysB6flcEx/.TOpfc;
sysB6olcEx = sysB6flcEx/.TOolc;
```

The expressions for the System B PFC, FLC and OLC models are

```
sysB6pfcEx//Length
```

4032

```
sysB6pfcEx//First
```

$1024\, r[\lambda\mathrm{A}, t]^6 r[\lambda\mathrm{AC}, t] r[\lambda\mathrm{C}, t] r[\lambda\mathrm{E}, t]^6 r[\lambda\mathrm{P}, t] r[\lambda\mathrm{S}, t]$

```
sysB6pfcEx//Last
```

$r[\lambda\mathrm{A}, t]^{12} r[\lambda\mathrm{AC}, t]^4 r[\lambda\mathrm{C}, t]^4 r[\lambda\mathrm{E}, t]^{48} r[\lambda\mathrm{P}, t]^4 r[\lambda\mathrm{S}, t]^4$

```
sysB6flcEx//Length
```

2664340

```
sysB6flcEx//First
```
1024 cAC1 cAC2 cAC3 cC1 cC2 cC3 cE1^6 cE2^6 cE3^6 cS1 cS2 cS3 $r[\lambda\text{A}, t]^6$ $r[\lambda\text{AC}, t] r[\lambda\text{C}, t] r[\lambda\text{E}, t]^6 r[\lambda\text{P}, t] r[\lambda\text{S}, t]$

```
sysB6flcEx//Last
```
67108864 cE1^{12} cE2^{12} cE3^{12} cS1 cS2 cS3 $r[\lambda\text{A}, t]^{12} r[\lambda\text{AC}, t]^4 r[\lambda\text{C}, t]^4$ $r[\lambda\text{E}, t]^{48} r[\lambda\text{P}, t]^4 r[\lambda\text{S}, t]^4$

```
sysB6olcEx//Length
```
98532

```
sysB6olcEx//First
```
1024 cAColc cColc cEolc^6 cSolc $r[\lambda\text{A}, t]^6 r[\lambda\text{AC}, t] r[\lambda\text{C}, t] r[\lambda\text{E}, t]^6$ $r[\lambda\text{P}, t] r[\lambda\text{S}, t]$

```
sysB6olcEx//Last
```
67108864 cEolc^{12} cSolc $r[\lambda\text{A}, t]^{12} r[\lambda\text{AC}, t]^4 r[\lambda\text{C}, t]^4 r[\lambda\text{E}, t]^{48} r[\lambda\text{P}, t]^4 r[\lambda\text{S}, t]^4$

Numerical results, computed from the non-expanded expressions, for System B reliability are

```
setNumeric;
times = {1, 2, 5, 10, 20};
Table[{t = times[[i]], N[1 - sysB6pfc, 6], N[1 - sysB6flc, 6],
N[1 - sysB6olc, 6]}, {i, 1, Length[times]}]//TableForm
setSymbolic;
```

1	2.42055×10^{-11}	2.20661×10^{-10}	1.72431×10^{-10}
2	9.92852×10^{-10}	1.38010×10^{-9}	1.28367×10^{-9}
5	7.27998×10^{-10}	1.94086×10^{-8}	1.91679×10^{-8}
10	4.43904×10^{-9}	1.51560×10^{-7}	1.51079×10^{-7}
20	4.19386×10^{-8}	1.20205×10^{-6}	1.20109×10^{-6}

6.6 Comparison of System A and System B Probability of Failure

The fully expanded polynomials for the System B CPM are large: the FLC polynomial has 2,644,340 terms, the OLC polynomial has 9,532 terms and the PFC polynomial has 4,032 terms. Although these polynomials are of considerable size, they are smaller than the System A polynomials: the PFC expression for System A has 12,271 terms. Recall that these are polynomials in **r[λ, t]** and that they are much smaller than the polynomials as they exist prior to the application of the i.i.d. substitution.

The probability of system failure for FLC, OLC and PFC versions of Systems A and B are shown numerically in Table 6.3, and graphical results are given in Figures 6.4 and 6.5. The FLC versions of Systems A and B are both shown graphically in Figure 6.6. The use of a single set of actuation computers, as with System B, is superior architecturally to the use of a set of computers for each control surface, as with System A. The System B approach would likely be the preferred choice from the perspective of cost, weight and reliability.

The reasons why System B is more reliable than System A may not be obvious. Consider the systems from the perspective of a total failure of their respective AC systems: for System A, the overall system will fail if any one of the six quad AC systems fails; the corresponding System B failure results from the failure of the one AC system. The System A failure represents a 1-out-of-6 failure, but for System B, the corresponding failure is a 1-out-of-1 failure. Using the failure rate of $\lambda_{AC} = 400$ fpmh and a mission time of $t = 20$ hours, the probability of failure for 1-out-of-6 AC systems (System A configuration), each made up of four AC components, is 5.881×10^{-8}, whereas the probability of a single 1-out-of-1 AC system (System B configuration) failure is 9.802×10^{-9}. System B is more likely than System A to fail as a result of an AC system failure.[2]

Systems A and B differ from the single actuation system only in the inclusion of five additional control surface actuation systems. Since the servo-loop electronics are not only highly reliable ($\lambda_E = 10$ fpmh) but also quad redundant, and since the actuators are both dual redundant and have a failure rate of only 2 fpmh, the series addition of five additional actuation systems does not greatly affect the overall system reliability.

---

[2] This analysis assumes that the AC components of System A and System B have identical failure rates. Even though the EL/ER components may well be housed in the same unit as the AC components, they are modeled separately because they are subject to independent failure. Thus, even though the AC units for the System B configuration may house six times as many servo-amplifiers (EL/ER components) as the System A configuration, the effect is taken into consideration by modeling the EL/ER component reliabilities separately.

Table 6.3 Probability of failure for Systems A and B

System A			
t (hrs)	PFC	FLC	OLC
1	2.43335×10^{-11}	3.65087×10^{-10}	2.36880×10^{-10}
2	1.01332×10^{-10}	2.05631×10^{-9}	1.79998×10^{-9}
5	8.07917×10^{-10}	2.79089×10^{-8}	2.72687×10^{-8}
10	5.71642×10^{-9}	2.17609×10^{-7}	2.16331×10^{-7}
20	6.23329×10^{-8}	1.73272×10^{-6}	1.73017×10^{-6}
System B			
1	2.42055×10^{-11}	2.20661×10^{-10}	1.72431×10^{-10}
2	9.92852×10^{-11}	1.38010×10^{-9}	1.28367×10^{-9}
5	7.27998×10^{-10}	1.94086×10^{-8}	1.91679×10^{-8}
10	4.43904×10^{-9}	1.51560×10^{-7}	1.51079×10^{-7}
20	4.19386×10^{-8}	1.20205×10^{-6}	1.20109×10^{-6}

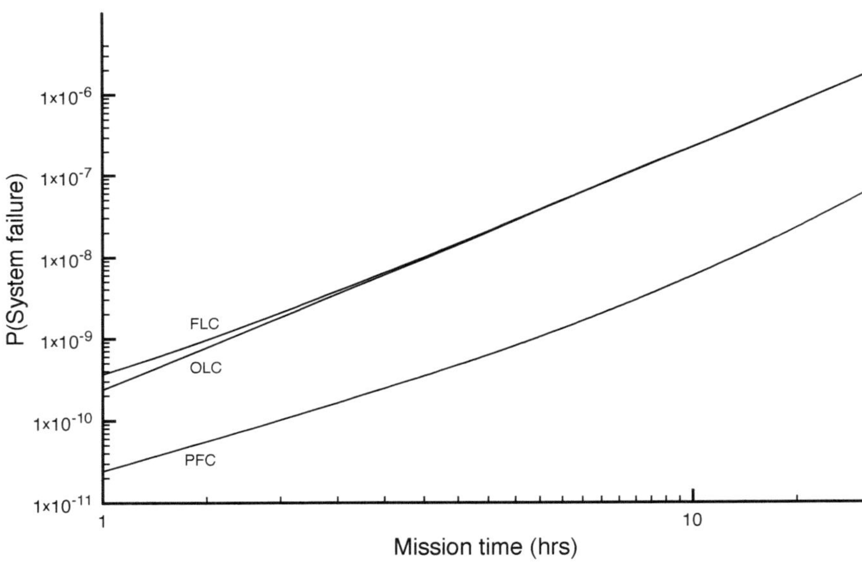

Fig. 6.4 Probability of failure for System A

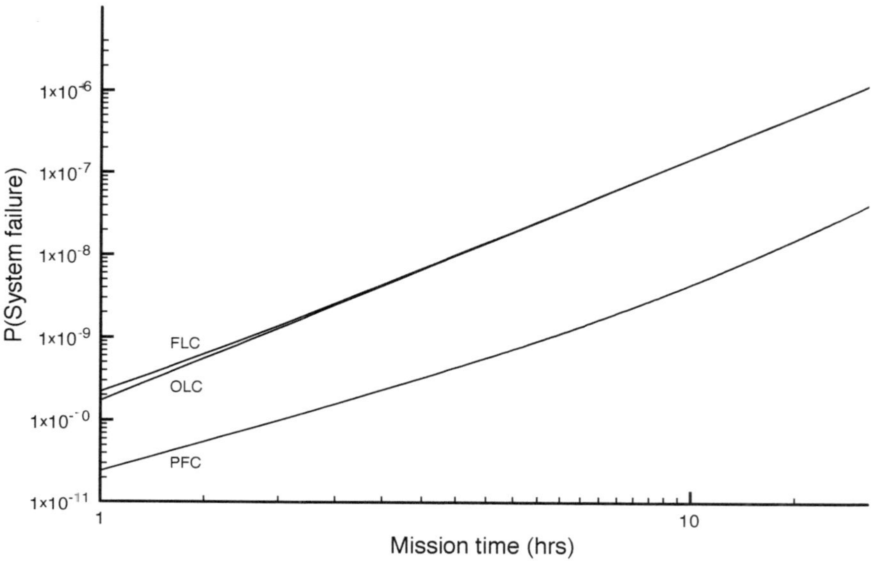

Fig. 6.5 Probability of failure for System B

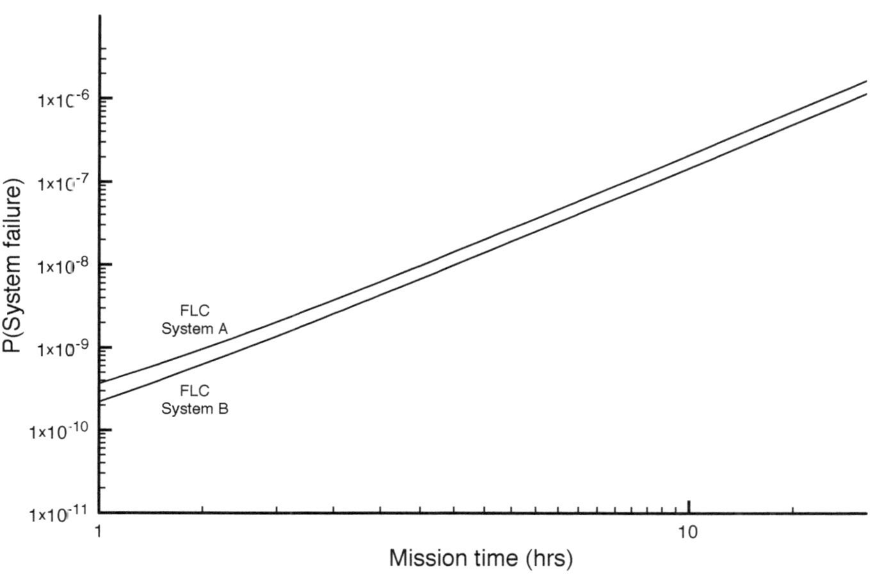

Fig. 6.6 Probability of failure for FLC Systems A and B

6.7 Comments on CPM

CPM techniques are a powerful method for the development of exact reliability models for complex multichannel systems. Although the use of BSV alone can model systems of only up to about two dozen independent components, the System A CPM covered in this chapter includes 106 components, and the System B model includes 80 components. Even though the examples in this section were rather simple, the approach outlined here can be extended to far more complex systems. For instance, the quad-redundant B-2 digital flight control system, which has 210 independent components, was modeled using CPM without the benefit of the *Mathematica*-based toolset using BSV. CPM requires breaking an overall system down into smaller (and hence more manageable) models, and BSV provides a convenient method for developing the required conditional component models.

In principle, systems of arbitrary size and complexity can be modeled using the combination of CPM and BSV discussed in this chapter. As long as the system can be characterized by a series of disjoint states with a complexity that makes each disjoint state small enough to be modeled using BSV or some other technique, the overall system model can be characterized as the sum of these disjoint products. In the examples of this chapter, the overall system was divided into a sequence of disjoint states with exactly i operational electrical buses or exactly j out of i operational channels. Determining the basis on which to predicate the conditionality, however, may not always be this straightforward; as was shown for System B, subtle points may make the correct selection of the conditionality basis difficult.

Although CPM techniques are a powerful approach, the determination of the appropriate conditional states can be challenging. In practice, the development of the required conditional probability equations is often a complex, tedious and (consequently) error-prone process. For systems of even modest complexity, a great deal of effort is required to validate the resulting models. Therefore, this approach has historically (especially prior to tools such as *Mathematica*-based BSV) been limited to the assessment of relatively mature designs. Furthermore, the resulting models have limits to their ability to support the analysis of architectural alternatives. A difference in system architecture, other than a simple change in the number of elements, nearly always has an effect on the conditional basis on which the model was built. Consequently, modeling a new system topology variation often requires redevelopment of the CPM.

The next chapter introduces an analysis method that uses binary decision diagram (BDD) techniques to model large systems in a fashion quite similar to the use of BSV but without the tight restrictions on the number of variables. Although BSV alone can reasonably handle up to about two dozen independent variables, BDD models can handle systems with hundreds of variables.

Chapter 7
Binary Decision Diagrams

...and the way is...

Abstract System reliability can be modeled through an analysis of the Boolean function that encodes the structure of a system. This relationship between the system structure and its corresponding Boolean function provides the foundation for the BSV analysis technique described in Chapter 3. The state-of-the-art technique for manipulating Boolean functions is the reduced order binary decision diagram (ROBDD)—often referred to simply as *BDD*. This chapter includes a brief overview of the BDD technique and provides BDD-based algorithms for the assessment of k-out-of-n:G structures for systems with both perfect and imperfect fault coverage.

7.1 Overview

The use of the reduced order binary decision diagram (ROBDD), referred to here as simply *BDD*, was initially developed by Bryant as a tool for validating VLSI circuit design [1]. An excellent description of a BDD package implementation is provided in [2].

BDDs were first used in reliability analysis in the early 1990s and, since then, have been demonstrated to be the most efficient technique for assessing fault trees and system reliability [3–6]. Although the space and time complexity for BDDs can be exponential in the worst case for most systems, the complexity is usually far better, and the use of BDDs has proven to be very effective in obtaining exact reliability results for large, complex multichannel systems.

This chapter demonstrates how the algorithms discussed in Chapter 4 are implemented using BDDs to yield the exact reliability assessments of PFC and IFC redundant systems. The table-based routines described in Section 4.7, when implemented using BDDs, constitute an extremely efficient approach for computing exact redundant system reliability results. The use of BDDs in the reliability assessment of redundant systems is discussed in [7].

7.1.1 Shannon Decomposition Theorem

The BDD representation is based on the Shannon decomposition theorem. Let F be a Boolean formula that depends on the variable v; the Shannon theorem can then be stated as

$$F = v \cdot F[v \leftarrow 1] + \overline{v} \cdot F[v \leftarrow 0] . \tag{7.1}$$

By assigning a specific order to the variables of a Boolean formula and recursively applying the Shannon decomposition in Equation (7.1), the truth table representing the Boolean formula can be depicted graphically as a binary decision tree, which is sometimes referred to as a *Shannon tree*.

7.1.2 Example

Consider the simple Boolean formula

$$\phi = (a \wedge b) \vee (\overline{a} \wedge c) . \tag{7.2}$$

The Shannon tree for ϕ in Equation (7.2) is shown below in Figure 7.1.

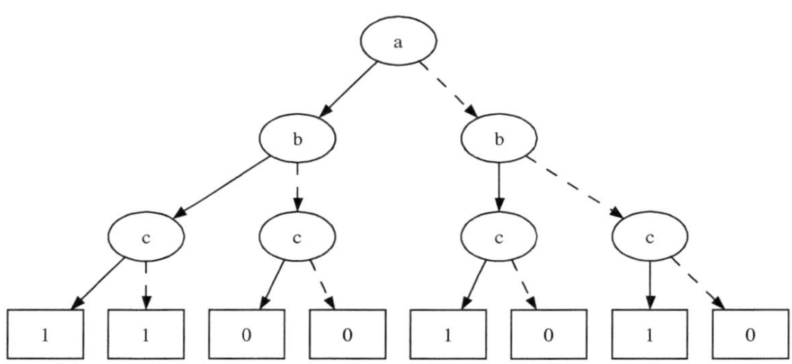

Fig. 7.1 Shannon tree representing $(a \wedge b) \vee (\overline{a} \wedge c)$

Each of the nodes in the tree represents a variable from ϕ. The nodes have two out edges: *then* (solid arrow) and *else* (broken arrow). The *then* edges represent the path taken if the variable is true, and the *else* edges represent the path taken if the variable is false. The tree has a set of terminal leaves, labeled 0 or 1, representing the value of the path obtained by traversing the corresponding branch. The value of the Boolean expression is the sum of the paths encoded in each of the branches leading to a 1 terminal leaf. The sum of the four paths leading to a 1 terminal leaf is

$$
\begin{aligned}
\phi &= abc + ab\overline{c} + \overline{a}bc + \overline{a}\overline{b}c \\
&= abc + ab(1-c) + (1-a)bc + (1-a)(1-b)c \\
&= ab + c - ac\,.
\end{aligned}
\tag{7.3}
$$

Alternatively, the value of ϕ could be computed as unity minus the sum of the paths leading to a 0 terminal leaf:

$$
\begin{aligned}
\phi &= 1 - (a\overline{b}c + a\overline{b}\overline{c} + \overline{a}b\overline{c} + \overline{a}\overline{b}\overline{c}) \\
&= 1 - (a(1-b)c + a(1-b)(1-c) + \\
&\qquad (1-a)b(1-c) + (1-a)(1-b)(1-c)) \\
&= ab + c - ac\,.
\end{aligned}
\tag{7.4}
$$

7.1.3 Reduction Rules

Although a Shannon tree, such as that shown in Figure 7.1, provides a graphical means of representing a truth table that encodes the value ϕ, the implementation is not efficient. The efficiency of the representation can be greatly improved both in time and memory, however, by applying the following reduction rules.

- Merge sub-trees that encode the same formula—only one is required.
- Delete unnecessary nodes—a node with both the *then* and *else* edges directed to the same node is not required ($\phi = v\phi + \overline{v}\phi$).

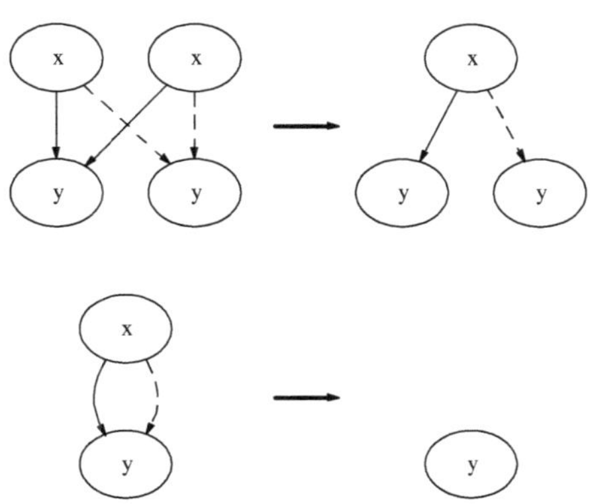

Fig. 7.2 BDD reduction rules

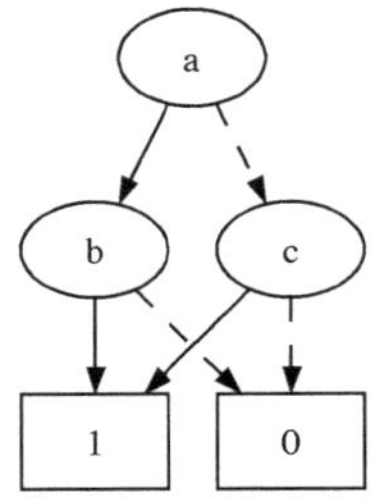

Fig. 7.3 BDD representing $(a \wedge b) \vee (\overline{a} \wedge c)$

These reduction rules are illustrated graphically in Figure 7.2. The exhaustive application of these rules to the Shannon tree shown in Figure 7.1 yields the reduced order binary decision diagram, or BDD, shown in Figure 7.3.

7.1.4 if-then-else *(ite) Function*

Logical operations, such as *and*, *or* and *not*, can be performed directly on a BDD. The implementation of BDD codes uses an *if-then-else* (ite) function to represent Boolean operations. This three-variable Boolean operator is defined as

$$\text{ite}(f, g, h) = f \cdot g + \overline{f} \cdot h\,. \tag{7.5}$$

The Shannon decomposition theorem, given in Equation (7.1), can be expressed in terms of the ite function:

$$f = \text{ite}(x_i, f\,|_{x_i=1}, f\,|_{x_i=0})\,. \tag{7.6}$$

Single-variable negation and all two-variable Boolean operations can be implemented using the ite function; those operations of interest in the formulation of reliability problems are shown in Table 7.1.[1]

7.1.5 BDD-Based k-out-of-n:G for PFC and IFC Systems

By applying the substitutions of Table 7.1 to the table-based PFC, ELC, FLC and OLC algorithms presented in Section 4.7, the BDD-based implementations shown in Figures 7.4 through 7.7 can be obtained. These table-based BDD routines were initially presented in [7].

[1] Recall from Section 3.4 that ⊗ implements the *and* function and ⊕ implements the *or* function (as opposed to the *exclusive or* function).

Table 7.1 Boolean constants and operations realized with an ite function

Boolean expression	BDD function	ite implementation
1, true	bddTRUE	1
0, false	bddFALSE	0
$f \otimes g$, $f \wedge g$	bddAND(f, g)	ite($f, g, 0$)
$f \oplus g$, $f \vee g$	bddOR(f, g)	ite($f, 1, g$)
$(1 - f)$, $\overline{f}$	bddNOT(f)	ite($f, 0, 1$)

```
Rbdd_PFC[k, p]
n ← length[p]
P ← array of length[n − k + 1]
P[1] ← bddTRUE
for i = 2 upto n − k + 1 do P[i] ← bddFALSE done
PFC ← bddFALSE
for i = 1 upto n do   for j = n − k downto 1 do
     p̄_i ← bddNOT(p_i)
     x ← bddAND(p_i, P[j + 1])
     y ← bddAND(p̄_i, P[j])
     P[j + 1] ← bddOR(x, y)
     if i == n then PFC ← bddOR(PFC, P[j + 1])
  done
  P[1] ← bddAND(P[1], p_i)
done
  PFC ← bddOR(PFC, P[1])
return PFC
```

Fig. 7.4 Table-based PFC implementation

```
Rbdd_ELC[k, p, c]
n ← length[p]
P ← array of length[n − k + 1]
P[1] ← bddTRUE
for i = 2 upto n − k + 1 do P[i] ← bddFALSE done
ELC ← bddFALSE
for i = 1 upto n do
  for j = n − k downto 1 do
     p̄_i ← bddNOT(p_i)
     x ← bddAND(p_i, P[j + 1])
     y ← bddAND(c_i, bddAND(p̄_i, P[j]))
     P[j + 1] ← bddOR(x, y)
     if i == n then ELC ← bddOR(ELC, P[j + 1])
  done
  P[1] ← bddAND(P[1], p_i)
done
  ELC ← bddOR(ELC, P[1])
return ELC
```

Fig. 7.5 Table-based ELC implementation

```
RbddFLC[k, p, c]
n ← length[p]
P ← array of length[n − k + 1]
P[1] ← bddTRUE
for i = 2 upto n − k + 1 do P[i] ← bddFALSE done
FLC ← bddFALSE
for i = 1 upto n do
  for j = n − k downto 1 do
    p̄i ← bddNOT(pi)
    x ← bddAND(pi, P[j + 1])
    y ← bddAND(cj, bddAND(p̄i, P[j]))
    P[j + 1] ← bddOR(x, y)
    if i == n then FLC ← bddOR(FLC, P[j + 1])
  done
  P[1] ← bddAND(P[1], pi)
done
  FLC ← bddOR(FLC, P[1])
return FLC
```

Fig. 7.6 Table-based FLC implementation

```
RbddOLC[k, p, c]

n ← length[p]
P ← array of length[n − k + 1]
P[1] ← bddTRUE
for i = 2 upto n − k + 1 do P[i] ← bddFALSE done
OLC ← bddFALSE
for i = 1 upto n do
  for j = n − k downto 1 do
    p̄i ← bddNOT(pi)
    x ← bddAND(pi, P[j + 1])
    if j == (n − 1) then
      P[j + 1] ← bddOR(x, bddAND(c, bddAND(p̄i, P[j])))
    else
      P[j + 1] ← bddOR(x, bddAND(p̄i, P[j]))
    if i == n then OLC ← bddOR(OLC, P[j + 1])
  done
  P[1] ← bddAND(P[1], pi)
done
  OLC ← bddOR(OLC, P[1])
return OLC
```

Fig. 7.7 Table-based OLC implementation

7.2 BDDs for *k*-out-of-*n*:G Systems

Graphical representations of BDDs for systems of modest size can be a helpful means of conveying the nature of system behavior in the face of component failures. In this section, 1-out-of-4:G systems consisting of redundant components $\mathbf{p} = \{p1, p2, p3, p4\}$ for PFC, ELC, FLC and OLC models are considered. The BDD for the PFC 1-out-of-4:G system is shown in Figure 7.8.

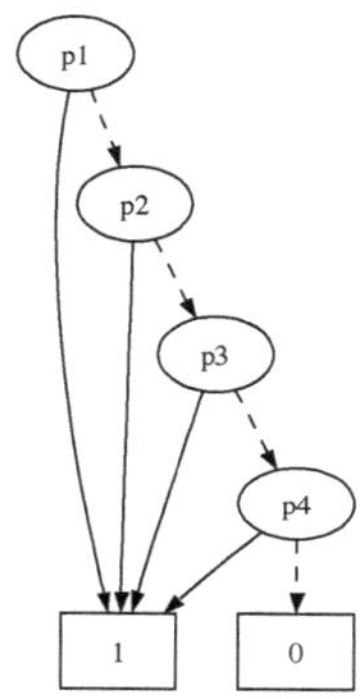

Fig. 7.8 BDD for a simple 1-out-of-4:G PFC system

The reliability of this system is the sum of all paths from the terminal node labeled 1 (true) to the top node, labeled p1, with the components that follow a broken-line connection being complemented. The reliability of the PFC 1-out-of-4:G system shown in Figure 7.8 is then

$$\begin{aligned} \mathrm{R_{PFC}} &= p1 + p2(1-p1) + p3(1-p2)(1-p1) \\ &\quad + p4(1-p3)(1-p2)(1-p1) \\ &= p1 + p2 - p1p2 + p3 - p1p3 - p2p3 \\ &\quad + p1p2p3\,. \end{aligned} \tag{7.7}$$

The sum of all paths taken in a similar manner from the terminal node labeled 0 (false) is the complement of the system reliability (or the probability of system failure), $1 - \mathrm{R_{PFC}}$:

$$\begin{aligned} \mathrm{R_{PFC}} &= 1 - (1-p4)(1-p3)(1-p2)(1-p1) \\ &= p1 + p2 - p1p2 + p3 - p1p3 - p2p3 \\ &\quad + p1p2p3\,. \end{aligned} \tag{7.8}$$

For the PFC system, the sum of the paths leading to node 0 has fewer terms: one term for this case versus four terms for the case of node 1. This characteristic, however, is not generally the case for IFC systems. A more compact presentation of the BDD omits the 0 terminal node, as shown in Figure 7.9.

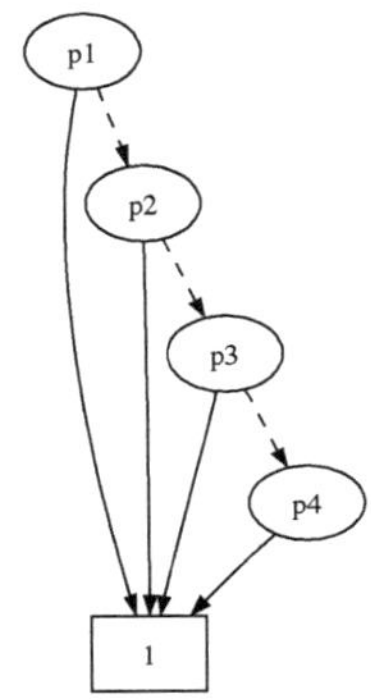

Fig. 7.9 BDD for a simple 1-out-of-4:G PFC system

An ELC system that also consists of four redundant components, **p**, and that has a coverage vector $\mathbf{c} = \{c1, c2, c3, c4\}$ (one coverage value for each redundant element) is shown in Figure 7.10. The BDDs for an FLC system with coverage vector $\mathbf{c} = \{c1, c2, c3\}$ (one coverage value for each sustainable failure) and for an OLC system with one-on-one coverage c are given in Figures 7.11 and 7.12, respectively.

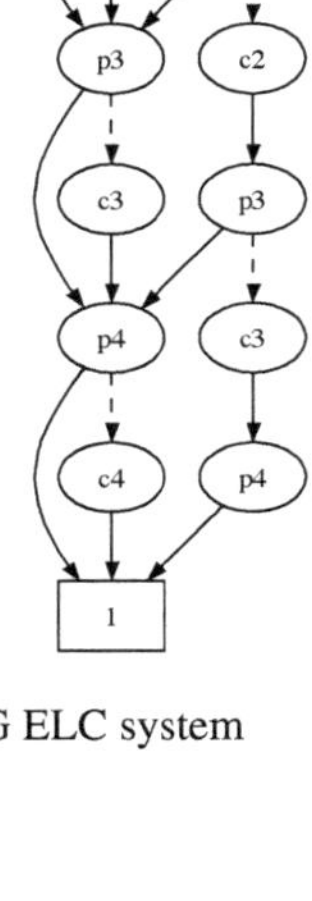

Fig. 7.10 BDD for a simple 1-out-of-4:G ELC system

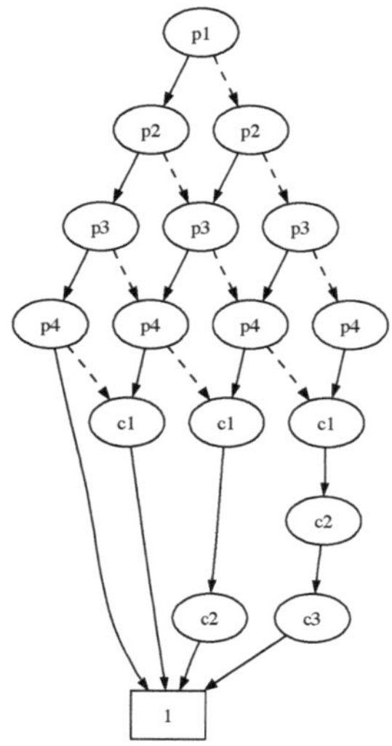

Fig. 7.11 BDD for a simple 1-out-of-4:G FLC system

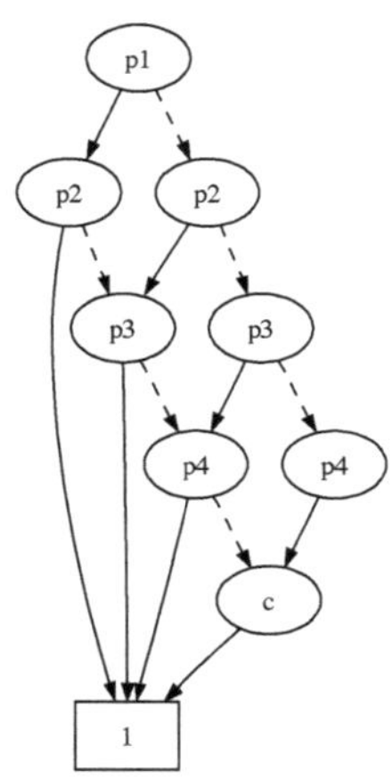

Fig. 7.12 BDD for a simple 1-out-of-4:G OLC system

System reliabilities that are computed from the BDDs for ELC, FLC and OLC k-out-of-n:G systems are algebraically equivalent to the values obtained from the combinatorial, recursive and table-based algorithms discussed in previous chapters.

7.3 BDD Comments and Observations

The use of BDDs represents the state-of-the-art approach to the determination of system reliability. Although BDD techniques are not immune from the potential for exponential growth (in both size and time) that results from the general reliability problem being NP-complete, the BDDs that implement most reliability problems, in practice, exhibit a complexity that is far superior to an exponential in the number of variables. The key to this substantial reduction in complexity is the correct ordering of the variables that define the problem. It has also been shown that the determination of an optimal variable ordering is itself NP-complete. Nevertheless, substantial empirical evidence suggests that the simple ordering procedure given below yields satisfactory performance for BDDs that describe redundant systems.

- Select variables in the sequence that they occur in a functional block diagram when moving from left to right (that is, from system input to system output).
- Place the coverage values of a redundant set immediately after the redundant set variables.

Using this simple variable-ordering approach, BDD analysis is quite successful for large, complex multichannel systems subject to either PFC or to IFC.

The following chapters discuss the use of the *FCASE* code, which employs the BDD approach summarized in this chapter, for the analysis of system reliability.

References

1. Bryant RE (1986) Graph-based algorithms for Boolean function manipulation. IEEE Trans Comp 35:677–691
2. Brace K, Ruddel R, Bryant R (1990), Efficient Implementation of a BDD Package, Proc 27th ACM/IEEE Des Autom Conf. IEEE 0738:40–45
3. Coudert O, Madre JC (1994) Metaprime: an interactive fault tree analyser. IEEE Trans Relia 43:121–127
4. Rauzy A (1993) New Algorithms for Fault Tree Analysis. Reliab Eng Syst Saf 40:203–211
5. Rauzy A (2001) Mathematical foundation of minimal cut sets. IEEE Trans Relia 50:389–396
6. Sinnamon RM and Andrews JD (1996) Quantitative fault tree analysis using Binary Decision Diagrams. Eur J Syst Autom 30:1051–1071
7. Myers A, Rauzy A (2008) Efficient Reliability Assessment of Redundant Systems Subject to Imperfect Fault Coverage Using Binary Decision Diagrams. IEEE Trans Relia 57:336–348

Chapter 8
FCASE Introduction

The cavalry is on the way!

Abstract This chapter provides a brief overview of the BDD-based *Flight Critical Aircraft System Evaluation* (*FCASE*) code for the determination of system reliability. A few simple example problems as well as an assessment of Systems A and B from Chapter 6 are used to illustrate *FCASE* operation. In subsequent chapters, greater detail is given regarding the full capabilities of *FCASE*.

8.1 Background

Several commercially available reliability programs are able to provide approximate solutions (and a smaller number are able to provide exact solutions) for the reliability of large, complex systems with PFC. Few, however, are able to provide straightforward solutions (approximate or exact) for systems subject to IFC of the type that is appropriate for multichannel voted systems requiring FLC or OLC modeling. Techniques for modeling ELC, for which the coverage of each redundant component has a fixed value that is independent of the fault sequence, have been covered in the literature [1,2] one such technique uses the SEA algorithm [3]. Techniques using SEA can be used in conjunction with PFC codes to obtain results for ELC systems, but they are not suitable for voted systems, which require either FLC or OLC.

The probability of failure for a system subject to IFC, even when the system has a high level of coverage, is significantly greater than the probability of failure for the corresponding "ideal" system with PFC. Furthermore, there are no straightforward techniques for estimating the IFC reliability from PFC results—particularly when the operational status of the redundant elements has upstream dependencies. To determine the reliability of systems subject to IFC, these systems must be modeled as such.

If a system is subject to IFC, as is the case with nearly all "real-world" redundant digital systems, the relative ranking of component contribution to system unreliability cannot be correctly evaluated by using a PFC model of the system. Additionally, as a result, system design trade-offs need to be evaluated using a model that correctly accommodates the effects of IFC.

The literature has generally not addressed (prior to [1–5]) the calculation of reliability for IFC systems that use voting to select among the redundant elements (that is, for FLC or OLC models).[1] Digital flight-control systems in modern military aircraft are typical examples of this kind of system design. In such systems, coverage is a function of the number of faults that the system has experienced, and these systems are modeled using either FLC or OLC. Aside from the *FCASE* code described in this chapter, the *Aralia* fault-tree code [6] is the only other currently available program that is capable of assessing the reliability of redundant systems that have a redundancy management approach subject to FLC (the type of imperfect fault coverage appropriate for voted systems). Both *FCASE* and *Aralia* can assess the full range of PFC, ELC, FLC and OLC systems.

Historically, reliability models for complex redundant systems subject to IFC had to be developed by deriving of a set of conditional probability equations that correctly describe the conditional interdependence of the system elements and that simultaneously accommodate the effects of imperfect fault coverage (as demonstrated in Chapter 6). These derivations tended to be difficult and tedious, and as a result, the process was prone to error. This difficulty meant that a great deal of the overall effort had to be dedicated to model verification. Consequently, fully verified models were generally not available until late in the overall system architecture design process; therefore, the myriad design trade-offs required to mature the system architecture were typically made on the basis of qualitative engineering judgments without the benefit of quantitative analysis. The requirement of extremely low probability of failure is the only reason for system redundancy, and this probability value can be modeled numerically and should be a fundamental element of system design.

This chapter provides an overview of the subset of the *FCASE* reliability analysis code which is useful in the illustration of the results achieved using the techniques and algorithms presented in earlier chapters. *FCASE* utilizes a BDD engine to perform the requisite system reliability assessment. Additional information regarding the use of *FCASE* can be found on `www.amyersconsulting.com`.

8.2 Simple System Example

For the description and explanation of each *FCASE* input file section, the inputs that are required to compute the reliability of the simple system shown in Figure 8.1 are used as examples.

For this example system to operate, component A must be operational, and at least one of the three B*i* components and at least one of the two C*i* components must also be operational. The failure rates for each of the element types are shown

[1] Combinatorial OLC models implemented in the context of conditional probability system models have been used to characterize digital fly-by-wire system reliability since the early 1980s. Nevertheless, these techniques were not published in the open literature at the time.

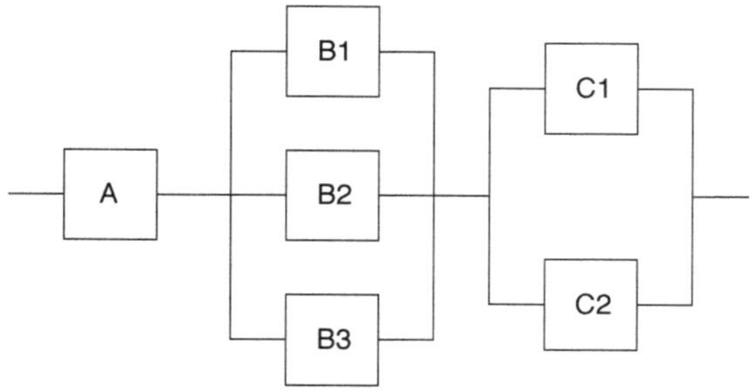

Fig. 8.1 System with simplex, triplex and duplex elements in series

Table 8.1 Component failure rates

Component	Failure rate (fpmh)
A	100
B1	1000
B2	1000
B3	1000
C1	500
C2	500

in Table 8.1. Note that the simple system depicted in Figure 8.1 can be easily characterized by any number of well-known techniques, including hand calculation. This system serves here as an example to illustrate the use of *FCASE*; *FCASE* modeling of more complex systems is treated in later sections.

8.2.1 FCASE *Input File Description*

The *FCASE* input file is delimited by four section headers that separate each type of processing required to solve the problem:

- `start VarDef`
- `start System`
- `start Results`
- `problem end`

Unlike *FCASE* commands, the section headers listed above have no terminating semicolon. All other *FCASE* statements are commands, which may include several contiguous lines, and are terminated with a semicolon (`;`). *FCASE* ignores "white space," and the user is encouraged to make use of this feature to format the code for improved readability.

The *FCASE* input file consists of three sections: a problem variable definition section (`start VarDef`), a problem definition section (`start System`) and a requested results definition section (`start Results`). The problem variable definition section defines the numerical values that describe the failure rates as well as the coverage factors applicable to the system. This section also defines the ordering of the variables that are used to build the BDD. The problem definition section describes the Boolean relationships between the components that define the system operation, and it is similar to the BSV models described in Chapter 5. The next section, `start Results`, defines the requested output of the analysis. The end of the input file is indicated with a `problem end` statement. Each of these input-file sections are further discussed below. Note that *FCASE* input-file statements are shown in the `typewriter` type face.

Two types of information are defined in the variable definition section, which is initiated with the `start VarDef` line, and these information types are defined in the following sequence: first, definition of numerical values for the individual component failure rates (that is, the lambda values) and any coverage factors that are applicable to the problem; second, definition of the variables that are used to define the system problem. The sequence in which these variables are defined is used to build the BDD. The variables can be scalars or vectors. Vectors are used when the system consists of redundant elements; for instance, a vector of length four is used to define a set of quad components.

8.2.1.1 Variable Definition Section (`start VarDef`)

The variable definition section starts with the following single line:

```
start VarDef
```

During execution, *FCASE* skips blank lines, comments and other white space in the input file; consequently, it is suggested that the user make liberal use of blank lines to facilitate readability.

Following this first line are the definitions of the applicable numerical values for component failure rates and for any coverage values. Each of these statements, like any *FCASE* command, is terminated with a semicolon (`;`). One statement is used for each of the required numerical definitions. For the present example, the following commands are used:

```
lA = 100e-6;
lB = 1000e-6;
lC = 500e-6;
```

The variables that are used to characterize the system and build the BDD are defined using the `bddVarDef` function. All component variables and coverage vectors (coverage scalars in the OLC case) must be defined in separate `bddVarDef` function calls.

Define a scalar variable for the A component, a vector of length three for the B components and a vector of length two for the C components. Building a BDD requires specification of the ordering of all variables that characterize the system; this variable ordering is taken to be the sequence in which they are defined.

```
bddVarDef(A, 1, exp, lA);
bddVarDef(B, 3, exp, lB);
bddVarDef(C, 2, exp, lC);
```

In the preceding code, the variable A is defined as a scalar (the second argument is 1) with an exponential survivor function (the third argument is `exp`) and with a failure rate `lA`. The B components are defined as a vector of length three with each of the components having a failure rate `lB`, and C is likewise defined as a vector of length two with failure rates `lC`.

In summary, the entire `VarDef` section for this example is

```
start VarDef

lA = 100e-6;
lB = 1000e-6;
lC = 500e-6;

bddVarDef(A, 1, exp, lA);
bddVarDef(B, 3, exp, lB);
bddVarDef(C, 2, exp, lC);
```

8.2.1.2 System Description Section (`start System`)

The system description section starts with the following single line:

```
start System
```

This section provides a Boolean characterization of the system and defines all of the interdependencies among the components that compose the system. This section uses two operators: the | operator, which represents a system reliability "OR" operation, and the & operator, which represents a system reliability "AND" operation. The | operator corresponds to the $\oplus$ operator, and the & operator corresponds to the $\otimes$ operator, as defined for BSV in Section 3.4. Both of these operators can operate on either scalar or vector operands. If two operands are vectors, then the operator is applied on a pairwise basis between each of the vector elements. Obviously, both vectors must be of equal length for this operation to be defined. Operations between a scalar and a vector are also defined; in this case, the scalar value operates on each of the vector elements.

For the current example problem, this section could be coded as follows:

```
start System

Bout = PFC(1, B);
Cout = PFC(1, C);
Sys_1 = A & Bout & Cout;
```

The PFC function provides the reliability of an *at least* k-out-of-n:G system subject to perfect fault coverage. *FCASE* also implements functions for *exactly* k-out-of-n:G systems subject to PFC as well as those subject to IFC. All of the *FCASE* k-out-of-n:G functions are described in Appendix E. The *FCASE* code above could be equivalently implemented as

```
Bout = A & (B1 | B2 | B3);
Sys_2 = Bout & (C1 | C2);
```

In this implementation, each of the vector components is identified by appending the vector element number to the vector name. For instance, C1 is the first element of C, and C2 is the second element. For this simple example, it is also easy to define the system using a single statement:

```
Sys_3 = A & (B1 | B2 | B3) & (C1 | C2);
```

The resulting BDD is the same for each of the alternatives; that is,

```
Sys_1 = Sys_2 = Sys_3
```

All three of these formulations could be expressed in a single *FCASE* input as follows:

```
start System

Bout = PFC(1, B);
Cout = PFC(1, C);
Sys_1 = A & Bout & Cout;

Bout = A & (B1 | B2 | B3);
Sys_2 = Bout & (C1 | C2);

Sys_3 = A & (B1 | B2 | B3) & (C1 | C2);
```

Results can be obtained for any variable defined in this section. The method for obtaining computed results is covered in the next section.

8.2.1.3 System Results Section (`start Results`)

The results section also starts with a single line:

```
start Results
```

The probability of failure as a function of time for any of the scalar variables that have been defined so far can be obtained as shown below. The following statements request that the probability of failure be tabulated for mission times starting at 1 hour and ending at 10 hours with increments of 1 hour.

```
repeat(Sys_1, 1, 10, 1);
repeat(Sys_2, 1, 10, 1);
repeat(Sys_3, 1, 10, 1);
```

FCASE can also calculate the contribution to system unreliability that is attributable to the failure rate of each component type. The following code creates a table that describes the effect of setting the failure rate of each component type, in turn, to zero. In other words, the code simulates the effect that perfect elements have on overall system unreliability at a mission time of 1 hour.

```
contribution(Sys_1, 1.0);
```

FCASE can also provide a symbolic expression for system reliability. The following code returns an algebraic expression for system reliability as a sum of disjoint products; each term represents the path through the BDD from the variable node to the "TRUE" terminal leaf.

```
path(Sys_1);
path(Sys_2);
path(Sys_3);
```

The end of problem statement, `problem end`, follows the last statement of the results section:

```
problem end
```

This statement also indicates the end of the *FCASE* input file.

8.2.1.4 Complete Input File for Simple System Example

FCASE input files can also include comments using the same syntax as that used in the C programming language (`/* This is a comment */`). The full *FCASE* input file for the example problem discussed above is then the following:

```
/* Very simple example system */

start VarDef

/* Define the component failure rates */

lA = 100e-6;
lB = 1000e-6;
lC = 500e-6;
```

```
/* Define the problem variables */

bddVarDef(A, 1, exp, lA);
bddVarDef(B, 3, exp, lB);
bddVarDef(C, 2, exp, lC);

start System

/* System Definition 1 */

Bout = PFC(1, B);
Cout = PFC(1, C);
Sys_1 = A & Bout & Cout;

/* System Definition 2 */

Bout = A & (B1 | B2 | B3);
Sys_2 = Bout & (C1 | C2);

/* System Definition 3 */

Sys_3 = A & (B1 | B2 | B3) & (C1 | C2);

start Results

/* Output the probability of system failure */
/* as a function of time, in each case from */
/* 1 to 20 hours in increments of 1 hour    */

repeat(Sys_1, 1, 10, 1);
repeat(Sys_2, 1, 10, 1);
repeat(Sys_3, 1, 10, 1);

/* Determine the sensitivity of system unreliability   */
/* to the failure rate of each of the component types. */
/* This analysis is done at a mission time of 1 hour.  */

contribution(Sys_1, 1.0);

/* Provide a symbolic result for each solution */

path(Sys_1);


path(Sys_2);
```

```
path(Sys_3);

problem end
```

The output that results from the execution of this input file is shown in the following section.

8.2.2 FCASE *Output File Description*

The first portion of the *FCASE* output file is simply an echo of the input file. This portion of an output file can be copied and pasted into a new input file as the starting point for a new problem.

In the output file shown below, the text presented using this style provides an annotation that is not part of the actual output, but has been included to provide a brief explanation of the different sections of the output file.

This portion of the *FCASE* output is an echo of the input file.

```
/* Very simple example system */

start VarDef

/* Define the component failure rates */

lA = 100e-6;
lB = 1000e-6;
lC = 500e-6;

/* Define the problem variables */

bddVarDef(A, 1, exp, lA);
bddVarDef(B, 3, exp, lB);
bddVarDef(C, 2, exp, lC);

start System

/* System Definition 1 */

Bout = PFC(1, B);
Cout = PFC(1, C);
```

```
Sys_1 = A & Bout & Cout;

/* System Definition 2 */

Bout = A & (B1 | B2 | B3);
Sys_2 = Bout & (C1 | C2);

/* System Definition 3 */

Sys_3 = A & (B1 | B2 | B3) & (C1 | C2);

start Results

/* Output the probability of system failure */
/* as a function of time, in each case from */
/* 1 to 10 hours in increments of 1 hour    */

repeat(Sys_1, 1, 10, 1);
repeat(Sys_2, 1, 10, 1);
repeat(Sys_3, 1, 10, 1);

/* Determine the sensitivity of system unreliability  */
/* to the failure rate of each of the component types.*/
/* This analysis is done at a mission time of 1 hour. */

contribution(Sys_1, 1.0);

/* Provide a symbolic result for each solution */

path(Sys_1);
path(Sys_2);
path(Sys_3);

problem end
```

The `problem end` statement indicates that the echo of the input file is complete; the rest of the output is execution output generated by *FCASE*.

The following initialization process will take a few seconds to complete.

```
Initializing BDD tables
```

The problem is now being defined and the BDD constructed.

```
Evaluate BDD node values for basic variables
```

Now the problem is being evaluated, or "solved"

```
Evaluating Problem
```

The following results are from the `repeat` requests for the probability of system failure as a function of time for 1 to 10 hours in increments of 1 hour.

```
Q(Sys_1) for t =  1.00 to  10.00

      1.00    1.00245849e-04
      2.00    2.00986777e-04
      3.00    3.02227829e-04
      4.00    4.03974015e-04
      5.00    5.06230306e-04
      6.00    6.09001636e-04
      7.00    7.12292905e-04
      8.00    8.16108975e-04
      9.00    9.20454671e-04
     10.00    1.02533478e-03

Q(Sys_2) for t =  1.00 to  10.00

      1.00    1.00245849e-04
      2.00    2.00986777e-04
      3.00    3.02227829e-04
      4.00    4.03974015e-04
      5.00    5.06230306e-04
      6.00    6.09001636e-04
      7.00    7.12292905e-04
      8.00    8.16108975e-04
      9.00    9.20454671e-04
     10.00    1.02533478e-03

Q(Sys_3) for t =  1.00 to  10.00

      1.00    1.00245849e-04
```

```
  2.00    2.00986777e-04
  3.00    3.02227829e-04
  4.00    4.03974015e-04
  5.00    5.06230306e-04
  6.00    6.09001636e-04
  7.00    7.12292905e-04
  8.00    8.16108975e-04
  9.00    9.20454671e-04
 10.00    1.02533478e-03
```

Note that the results for all three versions of the system description have identical numerical values, as expected.

The next output lists the contributions to system unreliability as a function of component failure rates.

```
Contribution to Unreliability (Sys_1) at t = 1.0
         Baseline Q = 1.0025e-04

Variable  Q w/ lambda = 0  Ratio to Baseline

lA             2.5087e-07       399.5872
lB             1.0024e-04         1.0000
lC             9.9996e-05         1.0025
```

The results above show that system unreliability is dominated by the failure rate of element A, which has a failure rate `lA`. If `lA` is set to 0, meaning that element A is made perfect, then system unreliability decreases by a factor of 399.5872. The results also show that improving the failure rates of the B and C elements has little effect on the overall system reliability.

The following are the symbolic results for the same three solution cases. As expected, the results for each of the cases are identical.

```
Symbolic Evaluation
```

In this case, the total number of paths through the BDD is 11; this number includes the paths that terminate at a TRUE leaf as well as the paths that terminate at a FALSE leaf.

```
Total Number of Paths = 11
```

This is a symbolic expression for all paths terminating at a TRUE leaf. The sum of the paths is equal to the system reliability.

```
R(Sys_1) =

A*B1*C1
+ A*B1*(1-C1)*C2
+ A*(1-B1)*B2*C1
+ A*(1-B1)*B2*(1-C1)*C2
+ A*(1-B1)*(1-B2)*B3*C1
+ A*(1-B1)*(1-B2)*B3*(1-C1)*C2
```

This expression shows that six paths terminate at TRUE.

```
Total Number of Success Paths = 6
```

The second solution:

```
Symbolic Evaluation
Total Number of Paths = 11

R(Sys_2) =

A*B1*C1
+ A*B1*(1-C1)*C2
+ A*(1-B1)*B2*C1
+ A*(1-B1)*B2*(1-C1)*C2
+ A*(1-B1)*(1-B2)*B3*C1
+ A*(1-B1)*(1-B2)*B3*(1-C1)*C2

Total Number of Success Paths = 6
```

The third solution:

```
Symbolic Evaluation
Total Number of Paths = 11
```

```
R(Sys_3) =

A*B1*C1
+ A*B1*(1-C1)*C2
+ A*(1-B1)*B2*C1
+ A*(1-B1)*B2*(1-C1)*C2
+ A*(1-B1)*(1-B2)*B3*C1
+ A*(1-B1)*(1-B2)*B3*(1-C1)*C2

Total Number of Success Paths = 6
```

In this example, the three variables, `Sys_1`, `Sys_2` and `Sys_3`, each represent systems analyzed using different formulations. For any given variable ordering, the resulting BDD for all equivalent system descriptions is unique to a tautology, and as shown above, the path expressions for each are identical.

The use of *FCASE* for the assessment of simple k-out-of-n:G systems subject to either PFC or IFC is illustrated in the next section.

8.3 *FCASE* 1-out-of-4:G PFC and IFC Examples

This section demonstrates the use of *FCASE* in the calculation of probability of system failure for the PFC, ELC, FLC and OLC models of the simple 1-out-of-4:G system shown in Figure 8.2. The redundant elements, Ai, have a failure rate of 1000 fpmh. The system BIT is assumed to have a coverage of 90%; as a result, all elements in the ELC coverage vector are equal to 0.9, and the OLC coverage value is equal to 0.9. The FLC coverage values are computed in the same manner as was discussed in Section 4.8 and are computed on the basis of a 30-millisecond fault detection window. *FCASE* provides the function `covCal` for computing the initial coverage values (that is, the coverage values that are prior to the one-on-one fault) for the FLC case, which is used in this example.

8.3.1 Simple 1-out-of-4:G System FCASE *Code and Results*

Since the *FCASE* output echoes the input file, only the output file is shown below.

```
start VarDef

lP = 1000.0e-6;

cELC = {0.9, 0.9, 0.9, 0.9};
```

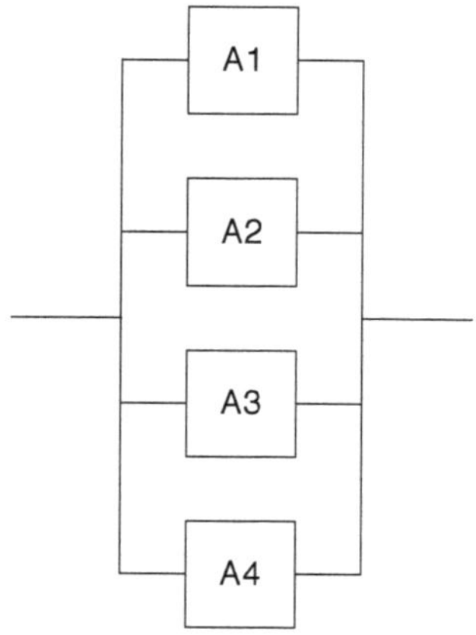

Fig. 8.2 Simple 1-out-of-4:G system

```
cFLC = {covCal(4, 1, lP, 30.0),
        covCal(4, 1, lP, 30.0), 0.9};
cOLC = 0.9;

bddVarDef(P, 4, exp, lP);
bddVarDef(cELC, cELC);
bddVarDef(cFLC, cFLC);
bddVarDef(cOLC, cOLC);

start System

sysPFC = PFC(1, P);
sysELC = ELC(1, P, cELC);
sysFLC = FLC(1, P, cFLC);
sysOLC = OLC(1, P, cOLC);

start Results

repeat(sysPFC, 1, 20, 1);
repeat(sysELC, 1, 20, 1);
repeat(sysFLC, 1, 20, 1);
repeat(sysOLC, 1, 20, 1);

problem end

Initializing BDD tables
```

```
Evaluate BDD node values for basic variables

Evaluating Problem

Q(sysPFC) for t =  1.00 to  20.00

      1.00    9.97979477e-13
      2.00    1.59361413e-11
      3.00    8.05155942e-11
      4.00    2.53960852e-10
      5.00    6.18783691e-10
      6.00    1.28054867e-09
      7.00    2.36763953e-09
      8.00    4.03102851e-09
      9.00    6.44404552e-09
     10.00    9.80215009e-09
     11.00    1.43227041e-08
     12.00    2.02447463e-08
     13.00    2.78287683e-08
     14.00    3.73564918e-08
     15.00    4.91306476e-08
     16.00    6.34747556e-08
     17.00    8.07329071e-08
     18.00    1.01269548e-07
     19.00    1.25469262e-07
     20.00    1.53736559e-07

Q(sysELC) for t =  1.00 to  20.00

      1.00    3.99740131e-04
      2.00    7.98961055e-04
      3.00    1.19766358e-03
      4.00    1.59584851e-03
      5.00    1.99351670e-03
      6.00    2.39066899e-03
      7.00    2.78730623e-03
      8.00    3.18342931e-03
      9.00    3.57903911e-03
     10.00    3.97413652e-03
     11.00    4.36872247e-03
     12.00    4.76279788e-03
     13.00    5.15636369e-03
     14.00    5.54942085e-03
     15.00    5.94197033e-03
     16.00    6.33401310e-03
```

```
        17.00    6.72555014e-03
        18.00    7.11658247e-03
        19.00    7.50711109e-03
        20.00    7.89713703e-03

Q(sysFLC) for t =  1.00 to  20.00

         1.00    4.99949082e-10
         2.00    3.39977690e-09
         3.00    1.10993778e-08
         4.00    2.59984807e-08
         5.00    5.04965653e-08
         6.00    8.69927813e-08
         7.00    1.37885867e-07
         8.00    2.05574071e-07
         9.00    2.92455069e-07
        10.00    4.00925890e-07
        11.00    5.33382834e-07
        12.00    6.92221399e-07
        13.00    8.79836197e-07
        14.00    1.09862089e-06
        15.00    1.35096809e-06
        16.00    1.63926934e-06
        17.00    1.96591495e-06
        18.00    2.33329401e-06
        19.00    2.74379429e-06
        20.00    3.19980212e-06

Q(sysOLC) for t =  1.00 to  20.00

         1.00    3.99999256e-10
         2.00    3.19997762e-09
         3.00    1.07998307e-08
         4.00    2.55992872e-08
         5.00    4.99978281e-08
         6.00    8.63946031e-08
         7.00    1.37188352e-07
         8.00    2.04777323e-07
         9.00    2.91559193e-07
        10.00    3.99930991e-07
        11.00    5.32289019e-07
        12.00    6.91028773e-07
        13.00    8.78544868e-07
        14.00    1.09723096e-06
        15.00    1.34947968e-06
```

```
16.00    1.63768255e-06
17.00    1.96422989e-06
18.00    2.33151080e-06
19.00    2.74191302e-06
20.00    3.19782292e-06
```

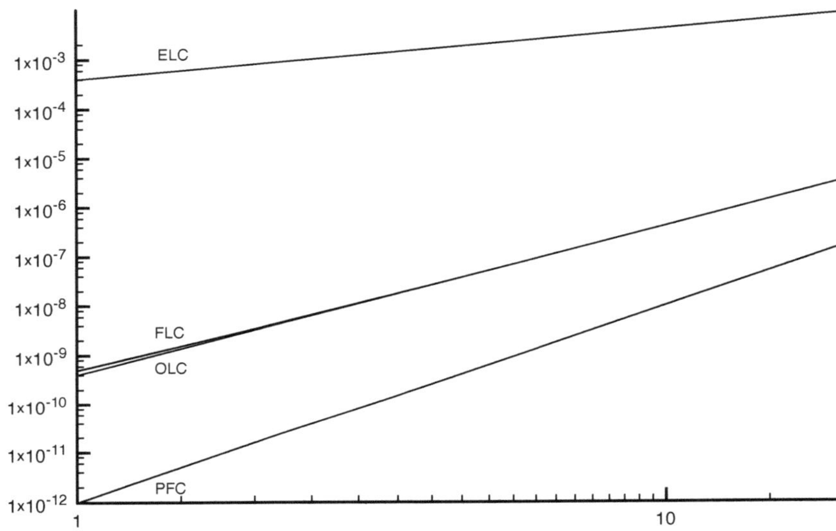

Fig. 8.3 Simple 1-out-of-4:G system probability of failure

The probability of system failure is shown in Figure 8.3. The data can be plotted by pasting the results from the *FCASE* output file into an Excel® spreadsheet and then generating a log-log chart. Alternatively, as was done here, another plotting package can be used.

8.4 Fly-by-Wire System with Six Control Surfaces System A and System B

In Chapter 6, CPM reliability models were developed for two versions of a simple fly-by-wire control system for an aircraft with six control surfaces: System A (shown in Figure 6.2) and System B (Figure 6.3). BSV techniques were used to derive the constituent elements of the CPM because the total number of elements in these systems was too large to be handled with a "direct" BSV model. *FCASE*, however, can easily model these systems in a direct fashion. The full *FCASE* file listings for both of the models are given in Appendix D.

The *FCASE* description for these systems is similar to that used in Chapter 5 for the derivation of system reliability models using BSV; the principal difference is the use of the & character to represent the function of the ⊗ operator and the | character to represent the function of the ⊕ operator. Another difference is the use of the individual functions PFC, PFCX, ELC, ELCX, FLC, FLCX, OLC and OLCX, rather than **RaL** and **ReX,** to represent the functions for computing k-out-of-n:G reliability for the PFC model and various IFC models.

The results obtained for the probability of system failure of System A and System B, for both a single surface (sSRFa) and the full system (sSys), are shown below.

8.4.1 FCASE *Results for System A*

```
Q(sSRFa) for t =  1.00 to  20.00 by incr 1.0000

         1.00     1.88736471e-10
         2.00     1.20550947e-09
         3.00     3.87111398e-09
         4.00     9.00711949e-09
         5.00     1.74358604e-08
         6.00     2.99804217e-08
         7.00     4.74646326e-08
         8.00     7.07130493e-08
         9.00     1.00550950e-07
        10.00     1.37804318e-07
        11.00     1.83299834e-07
        12.00     2.37864866e-07
        13.00     3.02327454e-07
        14.00     3.77516300e-07
        15.00     4.64260764e-07
        16.00     5.63390840e-07
        17.00     6.75737157e-07
        18.00     8.02130962e-07
        19.00     9.43404109e-07
        20.00     1.10038905e-06
```

```
Q(sSys) for t =  1.00 to  20.00 by incr 1.0000

     1.00    3.64957398e-10
     2.00    2.05605233e-09
     3.00    6.35409225e-09
     4.00    1.45426883e-08
     5.00    2.79082434e-08
     6.00    4.77399271e-08
     7.00    7.53296682e-08
     8.00    1.11972131e-07
     9.00    1.58964704e-07
    10.00    2.17607481e-07
    11.00    2.89203241e-07
    12.00    3.75057440e-07
    13.00    4.76478188e-07
    14.00    5.94776233e-07
    15.00    7.31264945e-07
    16.00    8.87260299e-07
    17.00    1.06408086e-06
    18.00    1.26304777e-06
    19.00    1.48548472e-06
    20.00    1.73271793e-06
```

8.4.2 FCASE *Results for System B*

```
Q(sSRFa) for t =  1.00 to  20.00 by incr 1.0000

     1.00    1.88736471e-10
     2.00    1.20550947e-09
     3.00    3.87111398e-09
     4.00    9.00711949e-09
     5.00    1.74358604e-08
     6.00    2.99804217e-08
     7.00    4.74646326e-08
     8.00    7.07130493e-08
     9.00    1.00550950e-07
    10.00    1.37804318e-07
    11.00    1.83299834e-07
    12.00    2.37864866e-07
    13.00    3.02327454e-07
    14.00    3.77516300e-07
    15.00    4.64260764e-07
    16.00    5.63390840e-07
```

```
       17.00    6.75737157e-07
       18.00    8.02130962e-07
       19.00    9.43404109e-07
       20.00    1.10038905e-06

  Q(sSys) for t =   1.00 to  20.00 by incr 1.0000

        1.00    2.20532148e-10
        2.00    1.37983625e-09
        3.00    4.36934400e-09
        4.00    1.00811505e-08
        5.00    1.94080044e-08
        6.00    3.32432952e-08
        7.00    5.24810445e-08
        8.00    7.80158909e-08
        9.00    1.10743079e-07
       10.00    1.51558451e-07
       11.00    2.01358428e-07
       12.00    2.61040009e-07
       13.00    3.31500751e-07
       14.00    4.13638755e-07
       15.00    5.08352669e-07
       16.00    6.16541659e-07
       17.00    7.39105407e-07
       18.00    8.76944103e-07
       19.00    1.03095842e-06
       20.00    1.20204953e-06
```

The results above provide the probability of failure for both a single-surface system, `Q(sSRFa)`, and for the full Systems A and B, `Q(sSys)`. When considering only a single control surface, System A and System B are equivalent, and as expected, the `Q(sSRFa)` values for both systems are identical. The `sSRFa` values correspond to the CPM-derived combined computer and actuation system values shown in Table 6.1. The System A and B results correspond to the CPM results in Table 6.3.

The numerical results that are obtained using *FCASE* are very close to those shown in Table 6.3, which were computed using the *Mathematica* CPM; the small variations are due to the differences in precision between the *Mathematica* and *FCASE* calculations. The *FCASE* BDD calculation uses standard double-precision arithmetic, and since computing the BDD-derived system reliability involves summing a large number of path expressions, some numerical differences can, in general, be expected. This is particularly the case for probabilities of failure less than about 10^{-10}. Such extremely small probabilities can be computed more accurately in *Mathematica* for two reasons: the simplification of the symbolic expression prior to numerical evaluation, and if necessary, the use of extended-precision arithmetic to compute a numerical result. Also note that the differences in numerical precision, in general, have a greater effect on FLC models and a lesser effect on OLC and PFC models.[2]

Although *FCASE* could also provide, in principle, a symbolic expression for these systems, the resulting polynomial expressions exceed the limit on the number of terms *FCASE* can provide (2×10^6). Nevertheless, *FCASE* could be used to generate the required symbolic expressions needed to develop a CPM for these systems, since they would be broken down into smaller subsystems. For most circumstances, however, the numerical *FCASE* solution is sufficient; the principle value of a CPM model is having an alternative solution technique during the validation process and for those circumstances in which an algebraic expression is required.

8.5 System B with Actuators in Series

The reliability of the system shown in Figure 6.3 is highly dependent on the redundancy of the actuators; for the surface to be operational, only one of the two actuators needs to be operational. Two critical system design features are required for actuator redundancy: first, the actuators must be sized such that in the event of a failure of the sister actuator, the remaining actuator can generate sufficient force to continue safe operation; second, the actuators must be designed to ensure that the failed actuator does not interfere with the operation of the good actuator. For hydraulic actuators, the second condition is facilitated by designing the actuator so that it can be placed in a bypass mode. The first condition is usually a simple matter of sizing, although this clearly has a system-wide effect on weight and power demand. For the simple system modeled in this example, the components that are required to implement a bypass mode have not been incorporated. Inclusion of these components would be necessary for a complete model.

If the actuators are sized so that both are required to operate the control surface or if the actuator system does not have the ability to bypass a failed actuator, then the actuators operate in series. Parallel actuators for one of the six surfaces are modeled as follows:

```
sSRFa = sAct1a | sAct2a;
```

The *FCASE* model can be easily modified to demonstrate the benefit of operating the actuators in series instead of in parallel. To model the system with the actuators in series, the corresponding *FCASE* line must be changed as follows for each of the six control surfaces:

```
sSRFa = sAct1a & sAct2a;
```

[2] The initial FLC coverage values are typically very close to unity and consequently are represented to about 10 significant digits, rather than the nominal 16 significant digits of standard double precision. The number of significant digits is usually not a concern with OLC coverage values.

The *FCASE* probability of system failure results for a modification of the FLC version of the system shown in Figure 6.3, with the actuators operating in series, are the following:

```
Q(sSys) for t =  1.00 to  20.00 by incr 1.0000

        1.00    2.40000982e-05
        2.00    4.80016129e-05
        3.00    7.20065346e-05
        4.00    9.60168515e-05
        5.00    1.20034549e-04
        6.00    1.44061612e-04
        7.00    1.68100021e-04
        8.00    1.92151756e-04
        9.00    2.16218792e-04
       10.00    2.40303104e-04
       11.00    2.64406665e-04
       12.00    2.88531444e-04
       13.00    3.12679408e-04
       14.00    3.36852523e-04
       15.00    3.61052750e-04
       16.00    3.85282049e-04
       17.00    4.09542379e-04
       18.00    4.33835693e-04
       19.00    4.58163946e-04
       20.00    4.82529087e-04
```

A comparison of these results with those for the case of parallel actuators demonstrates the degree to which the system reliability is dependent on actuator redundancy (that is, the degree to which it is dependent on having the actuators operate in parallel). The probability of failure for both systems is shown graphically in Figure 8.4. Even though this system has rather reliable actuators ($\lambda_A = 2$ fpmh), the system with series actuators is several orders of magnitude less reliable than the system with parallel actuators. Most likely, the series actuator system would not meet the probability of system failure requirements for an airborne system. The series actuator curve has a slope of approximately 1 dpd, which indicates that the resulting system is simplex in nature. This simplex character can also be expressed as an equivalent redundancy level (ERL) of approximately one. Designing a quad-redundant system with series actuators that have this level of reliability makes little sense. Indeed, if the balance of the system was dual redundant rather than quad redundant, then the probability of failure curves would have nearly the same ERL. As long as the system architecture uses non-redundant actuation, nothing can be done to significantly improve the probability of system failure, short of using actuators with an extraordinarily low failure rate.

It has already been noted that the simple system examples covered in this chapter do not include models of the components required for redundancy management for

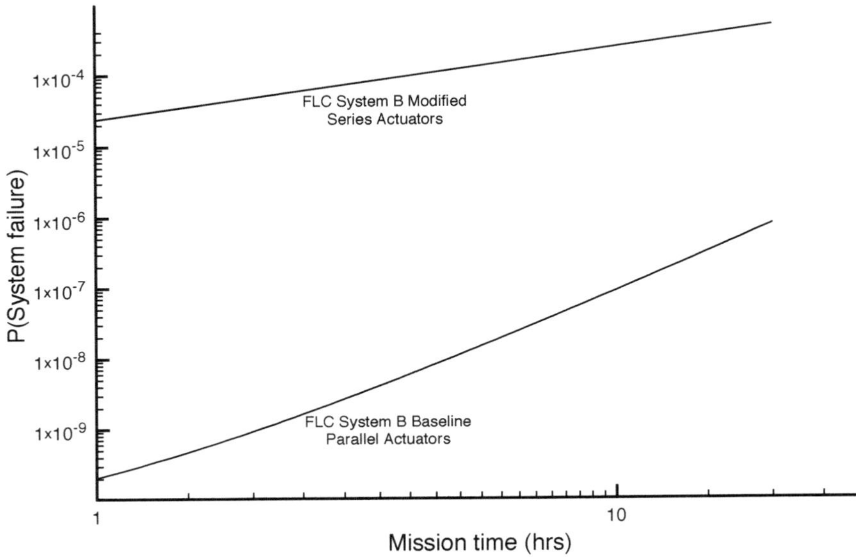

Fig. 8.4 System B with actuators in parallel and in series

the parallel actuators. Also, these system models do not include models of the power source for the actuators; instead, the system model makes the implicit assumption that if at least one of the servo-control elements is receiving power, then the actuator is also powered. The system analyzed in the next chapter provides a more complete model of a full digital flight control system, including the means for power delivery to the actuators.

References

1. Doyle SA, Dugan JB, Patterson–Hine FA (1995) A combinatorial approach to modeling imperfect coverage. IEEE Trans Relia 44:87–94
2. Amari SV, Dugan JB, Misra RB (1999) Optimal reliability of systems subject to imperfect fault coverage. IEEE Trans Relia 48:275–284
3. Amari SV, Dugan JB, Misra RB (1999) A separable method for incorporating imperfect fault-coverage in combinatorial models. IEEE Trans Relia 48:267–274
4. Myers AF (2007) *k*-out-of-*n*:G System Reliability With Imperfect Fault Coverage. IEEE Trans Relia 56:464–473
5. Myers A, Rauzy A (2008) Efficient Reliability Assessment of Redundant Systems Subject to Imperfect Fault Coverage Using Binary Decision Diagrams. IEEE Trans Relia 57:336–348
6. Rauzy A (2006) Aralia User's Manual. ARBoost Technologies

Chapter 9
Digital Fly-by-Wire System

Here they come!

Abstract This chapter discusses the use of the BDD-based *FCASE* code to assess the reliability of a hypothetical quadruple-redundant digital fly-by-wire (DFBW) system having an architecture typical of a military transport-class vehicle. The system, which is subject to imperfect fault coverage, has a probability of failure design requirement of less than or equal to 5×10^{-7} for a mission length of 10 hours. This example demonstrates the application of the previously discussed techniques in the assessment of a system of "real-world" size and complexity.

9.1 Quad-Channel DFBW System Description

A functional block diagram of a four-channel digital fly-by-wire control system is depicted in Figures 9.1 and 9.2 for a hypothetical four-engine military transport aircraft. This aircraft has a design mission duration $t_M = 10$ hours and is required to have a probability of loss of control (PLOC) less than or equal to 5×10^{-7} during a mission of duration t_M.[1] Since PLOC is a function of mission time, the term PLOC(t_M) is used to designate the probability of loss of control at $t = t_M$.

The primary means of redundancy management (RM) for this system is mid-value-select voting for detecting, isolating and reconfiguring the system in the event of failures among the redundant flight-critical components.[2] Since this RM task cannot be done with perfect certainty, the system is subject to imperfect fault coverage (IFC), and since the system uses voting, the appropriate IFC model is FLC. This

[1] PLOC is commonly understood to be the probability of loss of control, owing to the random failure of flight-critical components, during a mission of length t_M. Although this requirement would not normally include the loss of an aircraft because of engine failure, the engines *have* been included in this analysis. It is important to remember that PLOC only addresses issues of aircraft loss due to random component failure and does *not* address issues of aircraft loss due to exogenous circumstances such as structural failure, crew error, weather or battle damage (other specification requirements may, however, be applicable).

[2] Chapter 4 includes a more detailed discussion of RM.

chapter provides an FLC assessment of the system's PLOC over the period $t_M = 10$ hours. PFC and OLC results are also provided for purposes of comparison.

Most flight-critical components in this system are quad redundant, having a component associated with each of the four independent channels. For example, the system has four hydraulic pumps, HYD $\rightarrow$ (HYD 1, HYD 2, HYD 3, HYD 4), and four generators, GEN $\rightarrow$ (GEN 1, GEN 2, GEN 3, GEN 4); that is, the redundant set of generators are referred to as GEN. This convention is continued in the following description of the system.

The aircraft has four engines (ENG) that, in addition to providing the required thrust, power four independent electrical generators (GEN) and hydraulic pumps (HYD). The pilot-command transducers are also quad redundant. The pitch command inputs are designated PP, and the roll and yaw commands are designated PR and PY, respectively. The quad-redundant vehicle air-data information includes static pressure (PS) and total pressure (PT). The vehicle state information is provided by a redundant set of strap-down inertial reference units (INS). The analog PP, PR, PY, PS and PT transducer outputs are digitized by the four remote input/output units (RIO) and are transmitted to the four flight control computers (FCC) on a per-channel basis.

The redundant flight control computers execute identical software, on a frame-synchronous basis, at a 100 Hz rate (that is, a frame time of 10 ms). Each FCC communicates with each of the other three via a bidirectional data bus, referred to as a *cross-channel data link* (CCDL). By using the CCDL, each FCC transmits its own input data to each of the other three FCCs. Thus, each FCC has both its local data as well as the data from each of the other three FCCs; that is, each FCC has a full quad-redundant set of input data. Each FCC, using the RM approach outlined in Section 4.8, selects the inputs from among the redundant set that will be used to execute the vehicle's control laws and compute the control surface commands for processing by the actuation system.

A functional block diagram showing one of the vehicle's eight control surfaces is depicted in Figure 9.2. The actuation system is controlled by the set of actuation system computers (AC). Again, each of these computers executes identical software on a frame-synchronous basis at a 100 Hz frame rate. The control surface commands generated by each FCC are transmitted to the local AC. The actuation computers are also interconnected by a CCDL; this allows the ACs to select (vote) the control surface commands from among those computed by the FCCs in each of the operational channels.

The aircraft is controlled by eight control surfaces, all of which share the common architecture shown in Figure 9.2. Each control surface is controlled by two hydraulic actuators, both of which can be powered by either of two independent hydraulic systems. The actuators are controlled by a set of servo electronics (SEL). The SEL are physically housed within the ACs; that is, the AC 1 unit houses 16 sets of servo electronics—one set for each of the 16 actuators on the aircraft. The SEL i are shown as individual components because they are all subject to independent failure. In the event of an actuator (ACT) failure, the failed actuator is bypassed[3] by the actuation computers. If one of the actuators is bypassed, the control surface can be effectively controlled by the other actuator associated with that surface; thus, a control surface remains operational as long as at least one of its two actuators is operational. Control of the aircraft requires that all eight of its control surfaces be operational.

Note that since the flight control and actuation system computers each share their local input data with each of the other redundant computers via their CCDLs and since they perform identical RM, the outputs generated by each of the computers are identical as long as the computers are operational. In the event of either an FCC or AC failure, their respective outputs are removed from the voting set by the RM operating in the other redundant channels. Consequently, the overall flight control system is three-fail operational,[4] subject to FLC.

Figure 9.1 summarizes the failure rates for each of the components making up this quad-channel DFBW system. Coverage values for each of the "voted" components in this system are summarized in Figure 9.2.

Table 9.1 Component quantities and failure rates

Total system quantity	Component type	Failure rate, λ (fpmh)
4	ENG	100
4	GEN	200
4	HYD	350
4	PP	150
4	PR	150
4	PY	150
4	PS	100
4	PT	100
4	RIO	100
4	INS	250
4	FCC	200
4	AC	200
$4 \cdot 2 \cdot 8 = 64$	SEL	10
$2 \cdot 8 = 16$	ACT	5

Table 9.1 provides the total quantity of each component type along with its failure rate. The system consists of a total of 128 components, each subject to individual failure. Also note that most devices are subject to inoperability as a result of upstream failures; for instance, the flight control computers (FCC) are dependent on the availability of electrical power on their local electrical bus (BUS) and the actu-

[3] The actuators are designed to be placed in an unpowered bypass state in the event of a fault. Although a bypassed actuator cannot assist in controlling the surface, it does not restrict the motion of the other operational actuator.

[4] With the exception of the control surfaces, which have only duplex actuation.

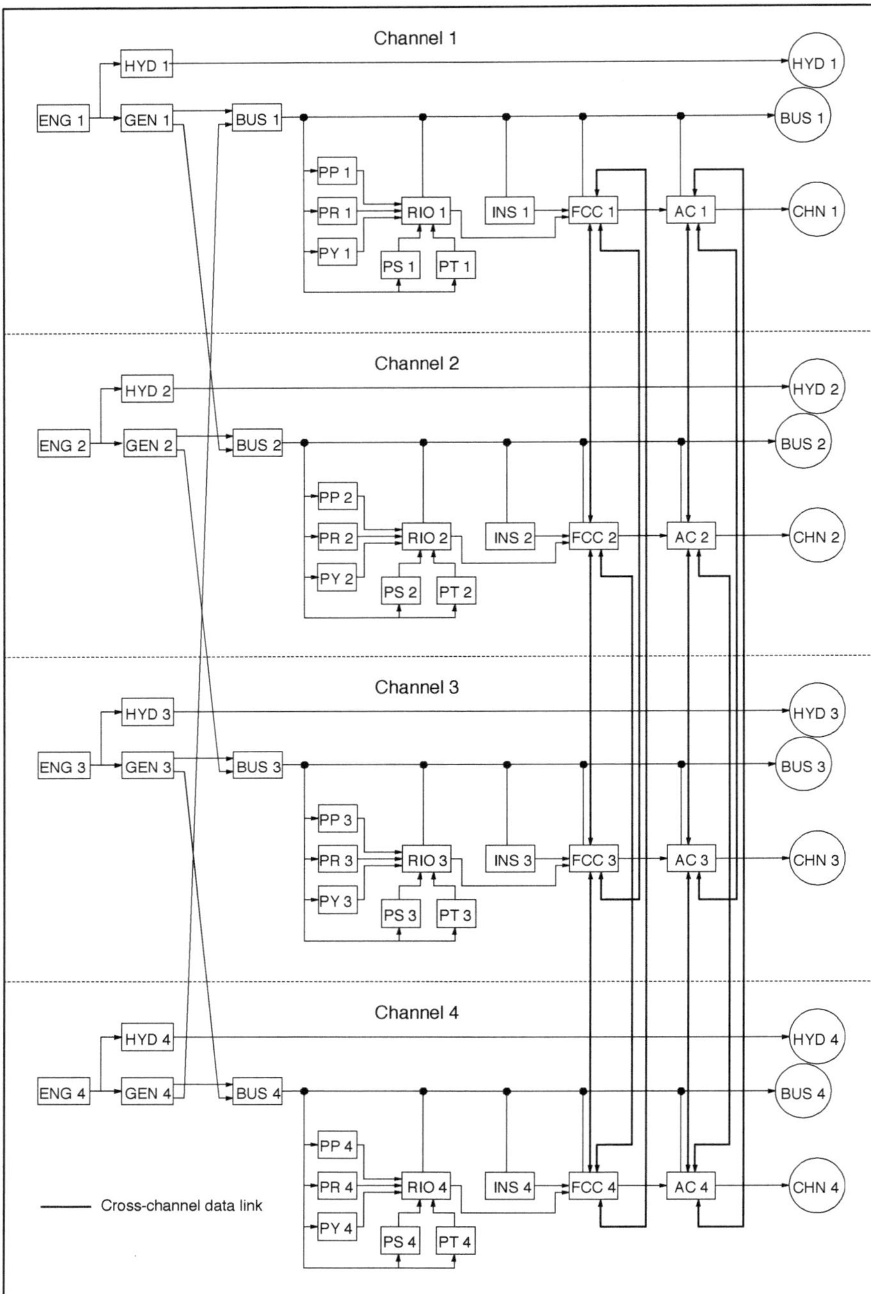

Fig. 9.1 Hypothetical quadruplex digital flight control system

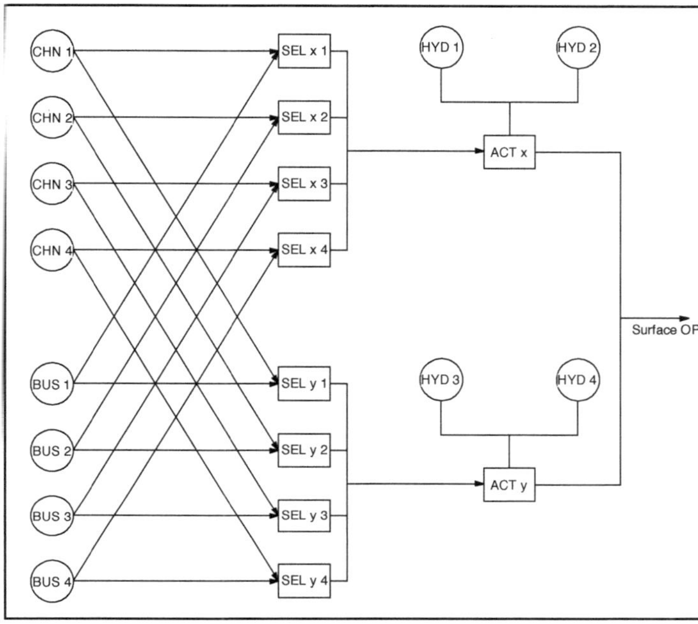

Fig. 9.2 Actuation system for one of eight control surfaces

Table 9.2 Coverage values for IFC components

Component type	OLC coverage c	FLC coverage		
		c_1	c_2	c_3
PP	0.95	0.99999999625	0.99999999750	0.95
PR	0.95	0.99999999625	0.99999999750	0.95
PY	0.95	0.99999999625	0.99999999750	0.95
PS	0.9	0.99999999750	0.99999999833	0.9
PT	0.9	0.99999999750	0.99999999833	0.9
INS	0.96	0.99999999375	0.99999999583	0.96
FCC	0.95	0.99999999500	0.99999999667	0.95
AC	0.95	0.99999999500	0.99999999667	0.95
SEL	0.9	0.99999999975	0.99999999983	0.9

ators (A) are dependent on the availability of hydraulic power from at least one of the two possible sources available to each.

The control computers are responsible for conducting the RM on all of their inputs: PP, PR, PY, PS, PT and INS. This RM process is subject to IFC, and the associated coverage values for the OLC and FLC models are shown in Table 9.2. The actuation system computers are responsible for the RM of the flight control computer outputs and of the servo-electronic components (SEL) shown in Figure 9.2. The actuation system computers are responsible for their own RM, and the coverage values for these elements are also provided in Table 9.2. The FLC coverage values were computed using the *FCASE* function `covCal`, which implements Equation (4.39) given in Section 4.8, and the OLC coverage values are the estimated BIT probabilities of the corresponding components.

Figure 9.2 depicts a functional diagram for one of the eight flight-critical control surfaces used by the vehicle. System operation requires all eight of the control surfaces to be operational. Each control surface uses two actuators (A). For the surface to be operational, at least one of the two actuators must be operational. Each of the actuators can be supplied hydraulic power from either of two hydraulic systems: (HYD 1, HYD 2) or (HYD 3, HYD 4). Each of the eight control surfaces must be operational for the aircraft to maintain safe flight.

9.2 *FCASE* Output File for Quad-Redundant DFBW System

The *FCASE* code and the FLC results for the analysis of the fly-by-wire system are shown below. The PFC and OLC input files can be determined in a similar fashion when the requisite changes to the coverage calculations and to the function calls (using either `PFC` or `OLC` as appropriate) have been made.

```
                                         FCASE
   Version: 07_10_09

    ************ Flight Critical Aircraft System Evaluation ***
   *********


         Run date: Sat Jul 11 08:52:12 2009

         Elapsed time: 0 seconds
   Input string to FCASE App


  /* Chapter 8 quad-redundant DFBW system    */
  /* This file determines FLC system results */

  /* Component failure rates set at nominal  */

  start VarDef

  /* Define component failure rates (lambda) */

  lENG = 100.0e-6;
  lGEN = 200.0e-6;
```

```
lHYD = 350.0e-6;
lPi = 150.0e-6;
lPr = 100.0e-6;
lRIO = 100.0e-6;
lINS = 250.0e-6;
lFCC = 200.0e-6;
lAC = 200.0e-6;
lSEL = 10.0e-6;
lACT = 5.0e-6;

/* Define the one-on-one coverage */

cPi = 0.95;
cPr = 0.9;
cINS = 0.96;
cFCC = 0.95;
cAC = 0.95;
cSEL = 0.9;

/* Determine the FLC coverage */

vcPP  = {covCal(4, 1, lPi, 30),  covCal(4, 2, lPi, 30), cPi};
vcPR  = {covCal(4, 1, lPi, 30),  covCal(4, 2, lPi, 30), cPi};
vcPY  = {covCal(4, 1, lPi, 30),  covCal(4, 2, lPi, 30), cPi};
vcPS  = {covCal(4, 1, lPr, 30),  covCal(4, 2, lPr, 30), cPr};
vcPT  = {covCal(4, 1, lPr, 30),  covCal(4, 2, lPr, 30), cPr};
vcINS = {covCal(4, 1, lINS, 30),
         covCal(4, 2, lINS, 30), cINS};
vcFCC = {covCal(4, 1, lFCC, 30),
         covCal(4, 2, lFCC, 30), cFCC};
vcAC  = {covCal(4, 1, lAC, 30),
         covCal(4, 2, lAC, 30), cAC};
vcSEL = {covCal(4, 1, lSEL, 30),
         covCal(4, 2, lSEL, 30), cSEL};

/* Define the BDD variables */

bddVarDef(ENG, 4, exp, lENG);
bddVarDef(GEN, 4, exp, lGEN);
bddVarDef(HYD, 4, exp, lHYD);
bddVarDef(RIO, 4, exp, lRIO);
bddVarDef(PP, 4, exp, lPi);
bddVarDef(cPP, vcPP);
bddVarDef(PR, 4, exp, lPi);
bddVarDef(cPR, vcPR);
```

```
bddVarDef(PY, 4, exp, lPi);
bddVarDef(cPY, vcPY);
bddVarDef(PS, 4, exp, lPr);
bddVarDef(cPS, vcPS);
bddVarDef(PT, 4, exp, lPr);
bddVarDef(cPT, vcPT);
bddVarDef(INS, 4, exp, lINS);
bddVarDef(cINS, vcINS);
bddVarDef(FCC, 4, exp, lFCC);
bddVarDef(cFCC, vcFCC);
bddVarDef(AC, 4, exp, lAC);
bddVarDef(cAC, vcAC);

/* Surface 1 */

bddVarDef(SELa,  4, exp,  lSEL);
bddVarDef(cSELa, vcSEL);
bddVarDef(ACTa,  1, exp,  lACT);

bddVarDef(SELb,  4, exp,  lSEL);
bddVarDef(cSELb, vcSEL);
bddVarDef(ACTb,  1, exp,  lACT);

/* Surface 2 */

bddVarDef(SELc,  4, exp,  lSEL);
bddVarDef(cSELc, vcSEL);
bddVarDef(ACTc,  1, exp,  lACT);

bddVarDef(SELd,  4, exp,  lSEL);
bddVarDef(cSELd, vcSEL);
bddVarDef(ACTd,  1, exp,  lACT);

/* Surface 3 */

bddVarDef(SELe,  4, exp,  lSEL);
bddVarDef(cSELe, vcSEL);
bddVarDef(ACTe,  1, exp,  lACT);

bddVarDef(SELf,  4, exp,  lSEL);
bddVarDef(cSELf, vcSEL);
bddVarDef(ACTf,  1, exp,  lACT);

/* Surface 4 */
```

```
bddVarDef(SELg,  4, exp,  lSEL);
bddVarDef(cSELg, vcSEL);
bddVarDef(ACTg,  1, exp,  lACT);

bddVarDef(SELh,  4, exp,  lSEL);
bddVarDef(cSELh, vcSEL);
bddVarDef(ACTh,  1, exp,  lACT);

/* Surface 5 */

bddVarDef(SELi,  4, exp,  lSEL);
bddVarDef(cSELi, vcSEL);
bddVarDef(ACTi,  1, exp,  lACT);

bddVarDef(SELj,  4, exp,  lSEL);
bddVarDef(cSELj, vcSEL);
bddVarDef(ACTj,  1, exp,  lACT);

/* Surface 6 */

bddVarDef(SELk,  4, exp,  lSEL);
bddVarDef(cSELk, vcSEL);
bddVarDef(ACTk,  1, exp,  lACT);

bddVarDef(SELl,  4, exp,  lSEL);
bddVarDef(cSELl, vcSEL);
bddVarDef(ACTl,  1, exp,  lACT);

/* Surface 7 */

bddVarDef(SELm,  4, exp,  lSEL);
bddVarDef(cSELm, vcSEL);
bddVarDef(ACTm,  1, exp,  lACT);

bddVarDef(SELn,  4, exp,  lSEL);
bddVarDef(cSELn, vcSEL);
bddVarDef(ACTn,  1, exp,  lACT);

/* Surface 8 */

bddVarDef(SELo,  4, exp,  lSEL);
bddVarDef(cSELo, vcSEL);
bddVarDef(ACTo,  1, exp,  lACT);

bddVarDef(SELp,  4, exp,  lSEL);
```

```
bddVarDef(cSELp, vcSEL);
bddVarDef(ACTp,  1, exp,  lACT);

start System

/* Problem definition */
/* Power system elements */

sENGGEN = ENG & GEN;
sENGHYD = ENG & HYD;

/* BUS is cross-strapped */

BUS =   {sENGGEN1 | sENGGEN4,
sENGGEN2 | sENGGEN1,
sENGGEN3 | sENGGEN2,
sENGGEN4 | sENGGEN3};

/* FCCvote is the voted output of the FCCs */
/* The output of the FCCs is dependent on the FCC coverage */

FCCvote = FLC(1, PP & BUS & RIO & FCC, cPP) &
 FLC(1, PR & BUS & RIO & FCC, cPR) &
 FLC(1, PY & BUS & RIO & FCC, cPY) &
 FLC(1, PS & BUS & RIO & FCC, cPS) &
 FLC(1, PT & BUS & RIO & FCC, cPT) &
 FLC(1, INS & BUS & FCC, cINS) &
 FLC(1, FCC & BUS, cFCC);

/* ACvote is the voted output of the ACs */

ACvote = FLC(1, AC & FCC & BUS & FCCvote, cAC);

/* Surface 1 */

srf1 = ACTa & FLC(1, ACvote & AC & SELa & BUS, cSELa) &
      (sENGHYD1 | sENGHYD2) |
       ACTb & FLC(1, ACvote & AC & SELb & BUS, cSELb) &
      (sENGHYD3 | sENGHYD4);

/* Surface 2 */

srf2 = ACTc & FLC(1, ACvote & AC & SELc & BUS, cSELc) &
      (sENGHYD1 | sENGHYD2) |
       ACTd & FLC(1, ACvote & AC & SELd & BUS, cSELd) &
```

```
      (sENGHYD3 | sENGHYD4);

/* Surface 3 */

srf3 = ACTe & FLC(1, ACvote & AC & SELe & BUS, cSELe) &
      (sENGHYD1 | sENGHYD2) |
       ACTf & FLC(1, ACvote & AC & SELf & BUS, cSELf) &
      (sENGHYD3 | sENGHYD4);

/* Surface 4 */

srf4 = ACTg & FLC(1, ACvote & AC & SELg & BUS, cSELg) &
      (sENGHYD1 | sENGHYD2) |
       ACTh & FLC(1, ACvote & AC & SELh & BUS, cSELh) &
      (sENGHYD3 | sENGHYD4);

/* Surface 5 */

srf5 = ACTi & FLC(1, ACvote & AC & SELi & BUS, cSELi) &
      (sENGHYD1 | sENGHYD2) |
       ACTj & FLC(1, ACvote & AC & SELj & BUS, cSELj) &
      (sENGHYD3 | sENGHYD4);

/* Surface 6 */

srf6 = ACTk & FLC(1, ACvote & AC & SELk & BUS, cSELk) &
      (sENGHYD1 | sENGHYD2) |
       ACTl & FLC(1, ACvote & AC & SELl & BUS, cSELl) &
      (sENGHYD3 | sENGHYD4);

/* Surface 7 */

srf7 = ACTm & FLC(1, ACvote & AC & SELm & BUS, cSELm) &
      (sENGHYD1 | sENGHYD2) |
       ACTn & FLC(1, ACvote & AC & SELn & BUS, cSELn) &
      (sENGHYD3 | sENGHYD4);

/* Surface 8 */

srf8 = ACTo & FLC(1, ACvote & AC & SELo & BUS, cSELo) &
      (sENGHYD1 | sENGHYD2) |
       ACTp & FLC(1, ACvote & AC & SELp & BUS, cSELp) &
      (sENGHYD3 | sENGHYD4);

/* System requires all 8 surfaces */
```

```
sSys = srf1 & srf2 & srf3 & srf4 & srf5 & srf6 & srf7 & srf8;

/* Generate probability of sSys as a function of t */

start Results

repeat(sSys, 1, 20, 1);

contribution(sSys, 1.0);
contribution(sSys, 10.0);

erl(sSys, 1.0);
erl(sSys, 10.0);

problem end


      Elapsed time: 0 seconds


Initializing BDD tables


Evaluate BDD node values for basic variables

elapsed time after evaluateBDDbasic: 2 seconds


Evaluating Problem

elapsed time after evaluateProblem: 22 seconds


 Q(sSys) for t =  1.00 to  20.00 by incr 1.0000

      1.00    3.98427180e-10
      2.00    2.08227069e-09
      3.00    5.94151606e-09
      4.00    1.28710921e-08
      5.00    2.37708434e-08
      6.00    3.95455099e-08
      7.00    6.11047015e-08
      8.00    8.93628705e-08
```

```
       9.00     1.25239297e-07
      10.00     1.69658050e-07
      11.00     2.23547976e-07
      12.00     2.87842667e-07
      13.00     3.63480439e-07
      14.00     4.51404308e-07
      15.00     5.52561960e-07
      16.00     6.67905734e-07
      17.00     7.98392592e-07
      18.00     9.44984099e-07
      19.00     1.10864639e-06
      20.00     1.29035014e-06

Contribution to Unreliability (sSys) at t = 1.0
        Baseline Q = 3.9843e-10

Variable  Q w/ lambda = 0  Ratio to Baseline

lENG          3.9191e-10        1.0166
lGEN          3.9829e-10        1.0003
lHYD          3.8299e-10        1.0403
lPi           3.5324e-10        1.1279
lPr           3.6682e-10        1.0862
lRIO          3.3679e-10        1.1830
lINS          3.7890e-10        1.0515
lFCC          2.6365e-10        1.5112
lAC           3.8085e-10        1.0461
lSEL          3.9843e-10        1.0000
lACT          1.8224e-10        2.1863

Contribution to Unreliability (sSys) at t = 10.0
        Baseline Q = 1.6966e-07

Variable  Q w/ lambda = 0  Ratio to Baseline

lENG          1.6219e-07        1.0460
lGEN          1.6843e-07        1.0073
lHYD          1.5393e-07        1.1022
lPi           1.3055e-07        1.2995
lPr           1.3991e-07        1.2126
lRIO          1.1409e-07        1.4871
lINS          1.5601e-07        1.0875
lFCC          5.9514e-08        2.8507
lAC           1.5581e-07        1.0889
lSEL          1.6965e-07        1.0001
```

```
1ACT          1.3353e-07          1.2705

Equivalent Redundancy Level (sSys) at t = 1.0
        ERL = 2.2418


Equivalent Redundancy Level (sSys) at t = 10.0
        ERL = 2.8881


                                   FCASE

 ************ Flight Critical Aircraft System Evaluation ***
*********



 ************         Normal End of Evaluation Run        ***
*********

elapsed time at problem end: 31 seconds
```

These results demonstrate that for a 10-hour mission, the system has a probability of failure of 1.7×10^{-7}— a value that meets the requirement of 5×10^{-7}. The `Contribution` results for a mission time of 10 hours indicate that the "weakest links" in the system are the flight control computers (FCC), which have an improvement ratio of 2.85. This result indicates that the overall system reliability would improve by a factor of 2.85 if the redundant element failure rate λ_{FCC} were zero. The next most significant contributors to system unreliability are the remote input/output units (RIO), which have an improvement ratio of 1.49. Since the improvement ratios for the balance of the system elements are only slightly over unity, it can be concluded that improvement of the reliability of these elements, while holding the flight control computer failure rate constant, provides only a modest improvement in the overall system reliability for mission times equal to 10 hours. Notice, however, that for a mission time of 1 hour, the actuators (ACT) are the greatest contributor to system unreliability, having an improvement ratio of 2.19. The unreliability for low mission times is caused by the dual redundancy of the actuators, whereas the balance of the system is quad redundant. Even highly reliable elements with reduced levels of redundancy can dominate the system reliability for low mission times. In this case, the actuator reliability and redundancy level are probably adequate, since the overall system meets its probability of failure requirement and since the actuators are not the primary source of unreliability at the designated mission time of 10 hours.

The ERL for a 10-hour mission time is 2.89; this value is consistent with what would be expected for a quad system that is subject to imperfect fault coverage (the best possible ERL for a quad IFC system is approximately three). The effect of the dual actuators can be seen at a 1-hour mission time: the ERL is reduced to 2.24.

9.3 Results for Quad DFBW System

Figure 9.3 depicts a plot of the probability of system failure curves for PFC, OLC and FLC models of the quad-redundant system shown in Figures 9.1 and 9.2.

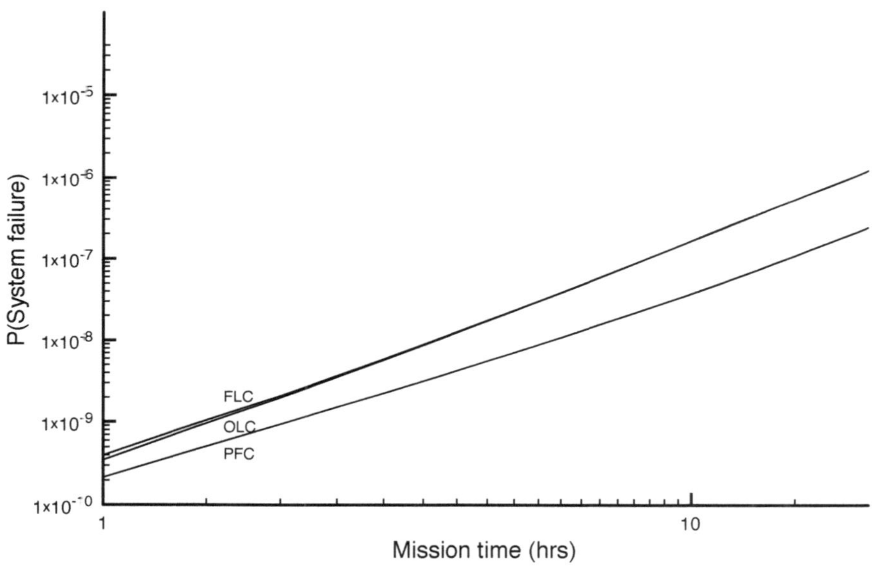

Fig. 9.3 Probability of failure for quadruplex digital flight control system

The PFC curve does not appear as a straight line; its slope for mission times less than 1 hour approaches an ERL of two, but for greater mission times, its slope increases, and the PFC ERL is approximately three for mission times greater than 20 hours. This result is a consequence of the system not being fully quad redundant: there are only two actuators per control surface, and the duplex actuators dominate the ERL for low mission times. If the actuators were also quad redundant, the general shape of the PFC, OLC and FLC curves would be the same as those for the simple 1-out-4:G system shown in Figure 8.3. For this system, OLC provides an excellent approximation of the FLC system.

This system also provides a clear example of the reason why one cannot correctly assess mission reliability by extrapolating the 1-hour probability of failure to higher mission times: the curves do not plot as straight lines on a log-log plot. As a consequence, such an extrapolation of the probability of failure to higher mission times would yield results that are overly optimistic.

The significant effect that IFC has on the probability of system failure is apparent in Figure 9.3. There is no "adjustment" that can be made to the PFC results to obtain correct IFC results: if the system is subject to IFC, then it must be modeled as such.

Calculation of these results on a 2.16 GHz Macintosh computer running OS X required approximately 2 seconds to create the BDD and "solve" the problem. An additional 31 seconds were required to compute the results for each of the requested times and other requested analyses using the BDD.

9.4 *FCASE* Output File for Triple-Redundant Fly-by-Wire System

The previous sections covered the analysis of a quad-redundant digital fly-by-wire system. The baseline FLC version of this quad system had a 1.7×10^{-7} probability of system failure for the design mission length of 10 hours—a reliability that easily meets the design requirement of 5×10^{-7}. It is useful to also consider the probability of failure for a triplex system. Modification of the *FCASE* file given above (for the quad system) to assess the performance of an otherwise identical triplex system is a straightforward task. Note that this triplex model has retained the quad power source components (ENG, GEN and HYD); consequently, the difference in system reliability is not due to a reduction in primary power redundancy.

The following is an *FCASE* output file for a triplex version of the digital fly-by-wire system analyzed in Section 9.1.

```
                                FCASE
   Version: 07_10_09

    ************ Flight Critical Aircraft System Evaluation ***
   *********


         Run date: Sat Jul 11 09:16:28 2009

         Elapsed time: 0 seconds
   Input string to FCASE App


  /* Chapter 8 triple-redundant DFBW system  */
  /* This file determines FLC system results */
```

```
/* Component failure rates set at nominal */

start VarDef

/* Define component failure rates (lambda) */

lENG = 100.0e-6;
lGEN = 200.0e-6;
lHYD = 350.0e-6;
lRIO = 100.0e-6;
lINS = 250.0e-6;
lFCC = 200.0e-6;
lAC = 200.0e-6;
lSEL = 10.0e-6;
lACT = 5.0e-6;

/* Define the one-on-one coverage */

cPi = 0.95;
cPr = 0.9;
cINS = 0.96;
cFCC = 0.95;
cAC = 0.95;
cSEL = 0.9;

/* Determine the FLC coverage */

vcPP  = {covCal(3, 1, lPi, 30), cPi};
vcPR  = {covCal(3, 1, lPi, 30), cPi};
vcPY  = {covCal(3, 1, lPi, 30), cPi};
vcPS  = {covCal(3, 1, lPr, 30), cPr};
vcPT  = {covCal(3, 1, lPr, 30), cPr};
vcINS = {covCal(3, 1, lINS, 30), cINS};
vcFCC = {covCal(3, 1, lFCC, 30), cFCC};
vcAC  = {covCal(3, 1, lAC, 30), cAC};
vcSEL = {covCal(3, 1, lSEL, 30), cSEL};

/* Define the BDD variables */

bddVarDef(ENG, 4, exp, lENG);
bddVarDef(GEN, 4, exp, lGEN);
bddVarDef(HYD, 4, exp, lHYD);
bddVarDef(RIO, 3, exp, lRIO);
bddVarDef(PP, 3, exp, lPi);
bddVarDef(cPP, vcPP);
```

```
bddVarDef(PR, 3, exp, lPi);
bddVarDef(cPR, vcPR);
bddVarDef(PY, 3, exp, lPi);
bddVarDef(cPY, vcPY);
bddVarDef(PS, 3, exp, lPr);
bddVarDef(cPS, vcPS);
bddVarDef(PT, 3, exp, lPr);
bddVarDef(cPT, vcPT);
bddVarDef(INS, 3, exp, lINS);
bddVarDef(cINS, vcINS);
bddVarDef(FCC, 3, exp, lFCC);
bddVarDef(cFCC, vcFCC);
bddVarDef(AC, 3, exp, lAC);
bddVarDef(cAC, vcAC);

/* Surface 1 */

bddVarDef(SELa,  3, exp,  lSEL);
bddVarDef(cSELa, vcSEL);
bddVarDef(ACTa,  1, exp,  lACT);

bddVarDef(SELb,  3, exp,  lSEL);
bddVarDef(cSELb, vcSEL);
bddVarDef(ACTb,  1, exp,  lACT);

/* Surface 2 */

bddVarDef(SELc,  3, exp,  lSEL);
bddVarDef(cSELc, vcSEL);
bddVarDef(ACTc,  1, exp,  lACT);

bddVarDef(SELd,  3, exp,  lSEL);
bddVarDef(cSELd, vcSEL);
bddVarDef(ACTd,  1, exp,  lACT);

/* Surface 3 */

bddVarDef(SELe,  3, exp,  lSEL);
bddVarDef(cSELe, vcSEL);
bddVarDef(ACTe,  1, exp,  lACT);

bddVarDef(SELf,  3, exp,  lSEL);
bddVarDef(cSELf, vcSEL);
bddVarDef(ACTf,  1, exp,  lACT);
```

```
/* Surface 4 */

bddVarDef(SELg,  3, exp,  lSEL);
bddVarDef(cSELg, vcSEL);
bddVarDef(ACTg,  1, exp,  lACT);

bddVarDef(SELh,  3, exp,  lSEL);
bddVarDef(cSELh, vcSEL);
bddVarDef(ACTh,  1, exp,  lACT);

/* Surface 5 */

bddVarDef(SELi,  3, exp,  lSEL);
bddVarDef(cSELi, vcSEL);
bddVarDef(ACTi,  1, exp,  lACT);

bddVarDef(SELj,  3, exp,  lSEL);
bddVarDef(cSELj, vcSEL);
bddVarDef(ACTj,  1, exp,  lACT);

/* Surface 6 */

bddVarDef(SELk,  3, exp,  lSEL);
bddVarDef(cSELk, vcSEL);
bddVarDef(ACTk,  1, exp,  lACT);

bddVarDef(SELl,  3, exp,  lSEL);
bddVarDef(cSELl, vcSEL);
bddVarDef(ACTl,  1, exp,  lACT);

/* Surface 7 */

bddVarDef(SELm,  3, exp,  lSEL);
bddVarDef(cSELm, vcSEL);
bddVarDef(ACTm,  1, exp,  lACT);

bddVarDef(SELn,  3, exp,  lSEL);
bddVarDef(cSELn, vcSEL);
bddVarDef(ACTn,  1, exp,  lACT);

/* Surface 8 */

bddVarDef(SELo,  3, exp,  lSEL);
bddVarDef(cSELo, vcSEL);
bddVarDef(ACTo,  1, exp,  lACT);
```

```
bddVarDef(SELp,  3, exp,  lSEL);
bddVarDef(cSELp, vcSEL);
bddVarDef(ACTp,  1, exp,  lACT);

start System

/* Problem definition */
/* Power system elements */

sENGGEN = ENG & GEN;
sENGHYD = ENG & HYD;

/* BUS is cross-strapped */

BUS =  {sENGGEN1 | sENGGEN2 | sENGGEN3 | sENGGEN4,
        sENGGEN1 | sENGGEN2 | sENGGEN3 | sENGGEN4,
        sENGGEN1 | sENGGEN2 | sENGGEN3 | sENGGEN4};

/* FCCvote is the voted output of the FCCs */
/* The output of the FCCs is dependent on the FCC coverage */

FCCvote = FLC(1, PP & BUS & RIO & FCC, cPP) &
 FLC(1, PR & BUS & RIO & FCC, cPR) &
 FLC(1, PY & BUS & RIO & FCC, cPY) &
 FLC(1, PS & BUS & RIO & FCC, cPS) &
 FLC(1, PT & BUS & RIO & FCC, cPT) &
 FLC(1, INS & BUS & FCC, cINS) &
 FLC(1, FCC & BUS, cFCC);

/* ACvote is the voted output of the ACs */

ACvote = FLC(1, AC & FCC & BUS & FCCvote, cAC);

/* Surface 1 */

srf1 = ACTa & FLC(1, ACvote & AC & SELa & BUS, cSELa) &
      (sENGHYD1 | sENGHYD2) |
       ACTb & FLC(1, ACvote & AC & SELb & BUS, cSELb) &
      (sENGHYD3 | sENGHYD4);

/* Surface 2 */

srf2 = ACTc & FLC(1, ACvote & AC & SELc & BUS, cSELc) &
      (sENGHYD1 | sENGHYD2) |
```

```
        ACTd & FLC(1, ACvote & AC & SELd & BUS, cSELd) &
       (sENGHYD3 | sENGHYD4);

/* Surface 3 */

srf3 = ACTe & FLC(1, ACvote & AC & SELe & BUS, cSELe) &
       (sENGHYD1 | sENGHYD2) |
        ACTf & FLC(1, ACvote & AC & SELf & BUS, cSELf) &
       (sENGHYD3 | sENGHYD4);

/* Surface 4 */

srf4 = ACTg & FLC(1, ACvote & AC & SELg & BUS, cSELg) &
       (sENGHYD1 | sENGHYD2) |
        ACTh & FLC(1, ACvote & AC & SELh & BUS, cSELh) &
       (sENGHYD3 | sENGHYD4);

/* Surface 5 */

srf5 = ACTi & FLC(1, ACvote & AC & SELi & BUS, cSELi) &
       (sENGHYD1 | sENGHYD2) |
        ACTj & FLC(1, ACvote & AC & SELj & BUS, cSELj) &
       (sENGHYD3 | sENGHYD4);

/* Surface 6 */

srf6 = ACTk & FLC(1, ACvote & AC & SELk & BUS, cSELk) &
       (sENGHYD1 | sENGHYD2) |
        ACTl & FLC(1, ACvote & AC & SELl & BUS, cSELl) &
       (sENGHYD3 | sENGHYD4);

/* Surface 7 */

srf7 = ACTm & FLC(1, ACvote & AC & SELm & BUS, cSELm) &
       (sENGHYD1 | sENGHYD2) |
        ACTn & FLC(1, ACvote & AC & SELn & BUS, cSELn) &
       (sENGHYD3 | sENGHYD4);

/* Surface 8 */

srf8 = ACTo & FLC(1, ACvote & AC & SELo & BUS, cSELo) &
       (sENGHYD1 | sENGHYD2) |
        ACTp & FLC(1, ACvote & AC & SELp & BUS, cSELp) &
       (sENGHYD3 | sENGHYD4);
```

```
/* System requires all 8 surfaces */

sSys = srf1 & srf2 & srf3 & srf4 & srf5 & srf6 & srf7 & srf8;

/* Generate probability of sSys as a function of t */

start Results

repeat(sSys, 1, 20, 1);

contribution(sSys, 1.0);
contribution(sSys, 10.0);

erl(sSys, 1.0);
erl(sSys, 10.0);

problem end



       Elapsed time: 0 seconds


Initializing BDD tables


Evaluate BDD node values for basic variables

elapsed time after evaluateBDDbasic: 2 seconds


Evaluating Problem

elapsed time after evaluateProblem: 2 seconds


 Q(sSys) for t =  1.00 to  20.00 by incr 1.0000

       1.00    2.32960685e-07
       2.00    9.32497234e-07
       3.00    2.09966553e-06
       4.00    3.73551711e-06
       5.00    5.84109909e-06
       6.00    8.41745428e-06
       7.00    1.14656211e-05
```

```
       8.00     1.49866336e-05
       9.00     1.89815215e-05
      10.00     2.34513102e-05
      11.00     2.83970208e-05
      12.00     3.38196700e-05
      13.00     3.97202703e-05
      14.00     4.60998297e-05
      15.00     5.29593520e-05
      16.00     6.02998367e-05
      17.00     6.81222791e-05
      18.00     7.64276700e-05
      19.00     8.52169961e-05
      20.00     9.44912398e-05

Contribution to Unreliability (sSys) at t = 1.0
        Baseline Q = 2.3296e-07

Variable  Q w/ lambda = 0  Ratio to Baseline

1ENG          2.3295e-07        1.0000
1GEN          2.3296e-07        1.0000
1HYD          2.3295e-07        1.0001
1Pi           1.8224e-07        1.2783
1Pr           1.9095e-07        1.2200
1RIO          1.6160e-07        1.4416
1INS          2.1341e-07        1.0916
1FCC          7.3373e-08        3.1750
1AC           2.0613e-07        1.1302
1SEL          2.3296e-07        1.0000
1ACT          2.3274e-07        1.0009

Contribution to Unreliability (sSys) at t = 10.0
        Baseline Q = 2.3451e-05

Variable  Q w/ lambda = 0  Ratio to Baseline

1ENG          2.3445e-05        1.0003
1GEN          2.3451e-05        1.0000
1HYD          2.3435e-05        1.0007
1Pi           1.8295e-05        1.2818
1Pr           1.9238e-05        1.2190
1RIO          1.6286e-05        1.4400
1INS          2.1452e-05        1.0932
1FCC          7.3928e-06        3.1722
1AC           2.0745e-05        1.1305
```

```
1SEL            2.3450e-05          1.0001
1ACT            2.3414e-05          1.0016

Equivalent Redundancy Level (sSys) at t = 1.0
        ERL = 2.0006


Equivalent Redundancy Level (sSys) at t = 10.0
        ERL = 2.0074


                                        FCASE

 ************ Flight Critical Aircraft System Evaluation ***
*********



 ************          Normal End of Evaluation Run        ***
*********

elapsed time at problem end: 9 seconds
```

These results for the triplex system show that PLOC(10) $= 2.35 \times 10^{-5}$, as opposed to 1.7×10^{-7} for the quad system. Thus, the change in redundancy from quadruplex to triplex has increased the probability of system failure by about two orders of magnitude, and the triplex system falls far short of meeting the design requirement of PLOC(10) $\leq 5 \times 10^{-7}$. The quadruplex and triplex systems are compared graphically in Figure 9.4.

Examination of the contribution analysis for the triplex system shows that the components that contribute the most to the system unreliability are the flight control computers (FCC). These results indicate that if the computers were replaced by perfect components, the probability of system failure would improve only by a factor of 3.17, whereas an improvement by a factor of 47 would be needed to meet the requirement of PLOC(10) $\leq 5 \times 10^{-7}$. This result suggests that selection of components with significantly greater reliability would be necessary for there to be any chance of meeting the design requirement.

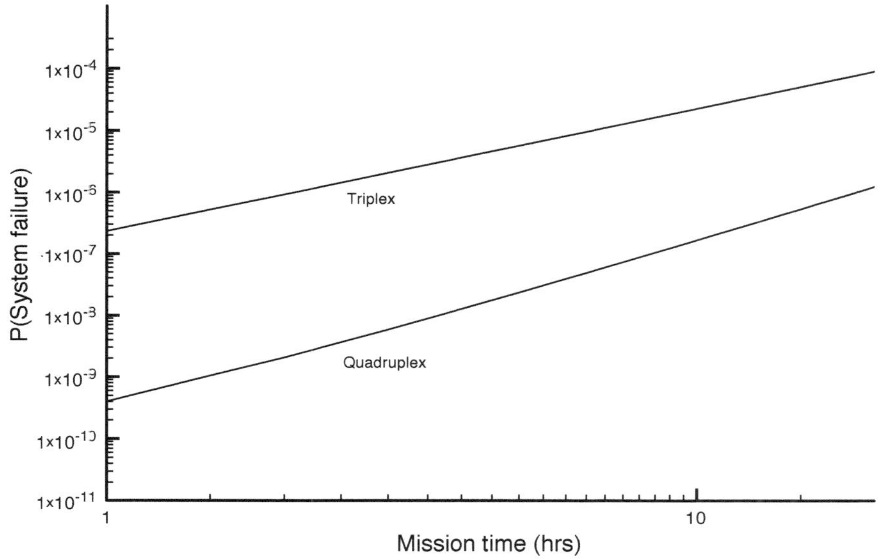

Fig. 9.4 Probability of failure for quadruplex and triplex digital flight control systems

Chapter 10
Limits on Achievable Reliability

There are some places you can't get to from here.

Abstract Although systems with perfect fault coverage can achieve arbitrarily high levels of system reliability by increasing the level of redundancy, n, this is not the case for systems with imperfect fault coverage. Attempts to enhance the reliability of IFC systems by arbitrarily increasing n will fail if $n > n_{\mathrm{opt}}$. IFC systems subject to either ELC or FLC have a finite n_{opt} value; beyond this value, the system reliability begins to decrease as the redundancy is further increased. This optimum redundancy level is different for the two types of IFC systems; it is, in general, greater for FLC systems than for ELC systems if both systems comprise components that have the same reliability and if the FLC system uses a voting-based redundancy management scheme with $n \geq 3$.

10.1 Introduction

The preceding chapters provide the means for computing the reliability of systems with either perfect or imperfect fault coverage. The effect of IFC on the probability of system failure for highly reliable systems is significant; for this reason, it is critical that the system designer, in the design process, appropriately account for the effects of IFC. There are two basic redundancy management architectures for redundant system RM: FLC for systems that use some form of mid-value-select voting to choose among the redundant elements, and ELC for systems that rely on built-in test as the primary redundancy management scheme.

Although systems with perfect fault coverage can achieve arbitrarily high levels of system reliability by increasing the level of redundancy, n, this is not the case for IFC systems. Attempts to enhance the reliability of IFC systems by arbitrarily increasing n will fail if $n > n_{\mathrm{opt}}$. IFC systems subject to either ELC or FLC have a finite n_{opt} value; beyond this value, the system reliability begins to decrease as the redundancy is further increased. This optimum redundancy level is different for the two types of IFC systems; it is greater for FLC systems than for ELC systems if both systems comprise components that have the same reliability.

10.2 IFC Models for i.i.d. *k*-out-of-*n*:G Systems

An FLC system is equivalent to the corresponding ELC system if all of the ELC and FLC coverage values are the same. Under these circumstances, the probability that a failure will be covered is identical for all failures for both the FLC and ELC models. There is, of course, no specific need for an FLC model if all coverage values are equal, since the model is only of interest when there is a difference between the coverage of initial and subsequent failures.

This chapter considers i.i.d. *k*-out-of-*n*:G systems using Equations (10.1) and (10.2), which were also given in Chapter 4. The results presented here also hold for the general case with unequal component reliabilities. Systems with non i.i.d. components can be assessed using the algorithms presented in Chapter 4.

The i.i.d. FLC function given below has a coverage vector $\mathbf{c}$, which contains $n-1$ non-identical values, but all coverage values for the ELC function are identical to c. The reliability of i.i.d. *k*-out-of-*n*:G ELC and FLC systems can be computed using the functions given in Equations (10.1) and (10.2), respectively. These equations correspond to Equation (4.18) from Chapter 4 and are based on the functions found in [1].

$$\mathrm{Riid}_{\mathrm{ELC}}(k, n, c) = \sum_{i=k}^{n} \binom{n}{i} p^i (q \cdot c)^{n-i} \tag{10.1}$$

$$\mathrm{Riid}_{\mathrm{FLC}}(k, n, \mathbf{c}) = \sum_{i=k}^{n} \binom{n}{i} p^i q^{n-i} \mathrm{cP}(i, n, \mathbf{c}) \tag{10.2}$$

Here, $\mathrm{cP}(k, n, \mathbf{c}) = \prod_{i=1}^{n-k} c_i$. Equations (10.1) and (10.2) are, of course, straightforward extensions of the well-known relationship for PFC *k*-out-of-*n*:G systems with i.i.d. elements (see [2]):

$$\mathrm{Riid}_{\mathrm{PFC}}(k, p) = \sum_{i=k}^{n} \binom{n}{i} p^i q^{n-i} . \tag{10.3}$$

As long as an FLC *k*-out-of-*n*:G system has at least three remaining operational components, it can use an MVS voting RM strategy to select among the active components. Nevertheless, once an FLC system has experienced $n - 2$ failures, it typically must rely on BIT (or a combination of BIT and system heuristics) to accomplish the RM tasks. Although an MVS vote can be accomplished with very high coverage, it can still be defeated if a nearly concurrent fault takes place before the system has had sufficient time to complete its fault detection and reconfiguration tasks. The time period required to complete these tasks is called the *fault detection window*, w. The coverage of these initial faults (also discussed in Section 4.8), which are prior to the $(n - 2)$th fault, can be estimated as

$$c_i = \mathrm{e}^{-(n-i)\lambda w} . \tag{10.4}$$

In Equation (10.4), i is the fault number in sequence: $i = 1$ corresponds to the first fault, $i = 2$ corresponds to the second fault and so on.

As discussed above, c_{n-1} can be estimated as the BIT (or some combination of BIT and system heuristics) coverage. An FLC k-out-of-n:G system has a total of $n - 1$ coverage values.

10.3 Optimum Reliability for IFC 1-out-of-n:G Systems

Redundant systems subject to imperfect fault coverage have an optimum level of redundancy; above this level, additional redundancy results in a decrease in system reliability. This section discusses the optimum level of redundancy for both ELC and FLC systems.

10.3.1 Optimum ELC 1-out-of-n:G Systems

The value of n_{opt} for an ELC 1-out-of-n:G system is easily determined using Equation (10.1) by simply increasing the value of n until the resulting reliability begins to decrease. This maximum reliability corresponds to the optimum redundancy level, n_{opt}.

Optimum redundancy levels and resulting system reliabilities for i.i.d. ELC 1-out-of-n:G systems are given in Table 10.1 for a range of i.i.d. component reliabilities, p, and coverage values, c. A graphical depiction of n_{opt} as a function of c and p is given in Figure 10.1.

10.3.2 Optimum FLC 1-out-of-n:G Systems

Equation (10.2) can be used to determine the optimum redundancy level for 1-out-of-n:G FLC systems using the same approach as that used above for ELC systems.

Reliability assessment of an FLC 1-out-of-n:G system requires estimation of $n-1$ coverage values. To assess the ELC and FLC systems on a comparable basis, the $(n-1)$th coverage value is considered to be equal to the BIT coverage, as was done above for each of the ELC system coverage values. The initial coverage values, $c_1, \ldots, c_{n-2}$, are estimated using Equation (10.4) using a fault detection window (w) of 30 milliseconds (three successive 10-millisecond frames). A more detailed discussion of the estimation of ELC coverage is given in Section 4.8.

Table 10.2 summarizes the optimum redundancy level, n_{opt}, and the lowest achievable probability of system failure, Q_{opt}, for FLC 1-out-of-n:G systems over the same range of i.i.d. component reliabilities and BIT coverage values as are used for ELC systems in Table 10.1. Graphical results for FLC 1-out-of-n:G systems

Table 10.1 ELC optimum redundancy level, n_{opt}, and probability of system failure, Q_{opt}

Component reliability, p	ELC coverage, c			
	0.9999	0.999	0.99	0.9
0.999999	2 2.01000×10^{-10}	2 2.00100×10^{-9}	2 2.00010×10^{-8}	2 2.00001×10^{-7}
0.99999	2 2.09998×10^{-9}	2 2.00998×10^{-8}	2 2.00098×10^{-7}	2 2.00008×10^{-6}
0.9999	2 2.99980×10^{-8}	2 2.09980×10^{-7}	2 2.00980×10^{-6}	2 2.00080×10^{-5}
0.999	3 3.01000×10^{-7}	2 2.99800×10^{-6}	2 2.09800×10^{-5}	2 2.00800×10^{-4}
0.99	3 3.99970×10^{-6}	3 3.09997×10^{-5}	2 2.98000×10^{-4}	2 2.08000×10^{-3}
0.9	5 5.99940×10^{-5}	4 4.99541×10^{-4}	3 3.96730×10^{-3}	2 2.80000×10^{-2}
0.8	7 1.52783×10^{-4}	6 1.26302×10^{-3}	4 9.51299×10^{-3}	3 6.46400×10^{-2}
0.7	9 2.89633×10^{-3}	7 2.31528×10^{-2}	5 1.72212×10^{-2}	3 1.07010×10^{-1}

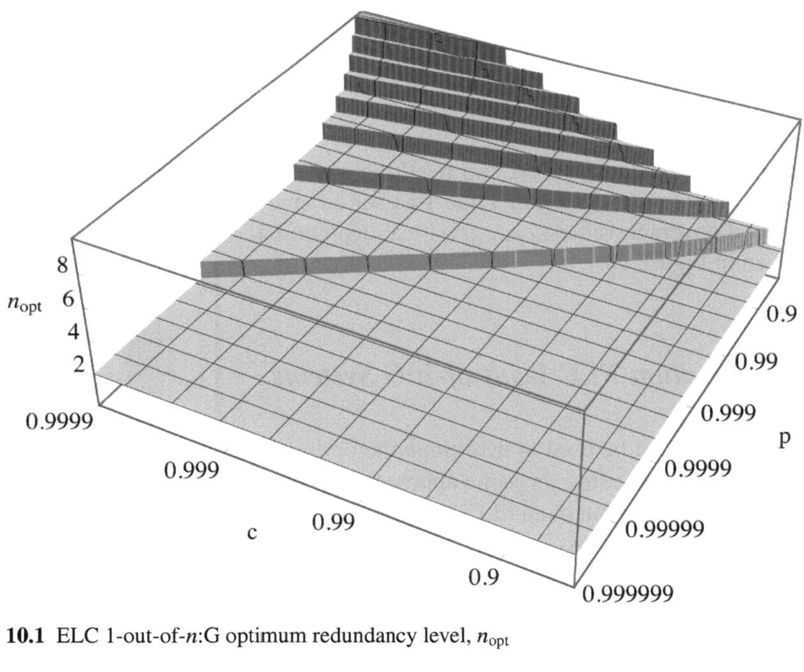

Fig. 10.1 ELC 1-out-of-n:G optimum redundancy level, n_{opt}

showing n_{opt} as a function of coverage (c_{n-1}) and i.i.d. component reliability (p) are given in Figure 10.2.

Table 10.2 FLC optimum redundancy level, n_{opt}, and probability of system failure, Q_{opt}

Component reliability, p	FLC coverage, c_{n-1}			
	0.9999	0.999	0.99	0.9
0.999999	4 1.00000×10^{-16}	4 1.00004×10^{-16}	4 1.00040×10^{-16}	4 1.00400×10^{-16}
0.99999	4 1.00004×10^{-14}	4 1.00040×10^{-14}	4 1.00400×10^{-14}	4 1.04000×10^{-14}
0.9999	4 1.00050×10^{-12}	4 1.00410×10^{-12}	4 1.04010×10^{-12}	4 1.40006×10^{-12}
0.999	4 1.01400×10^{-10}	4 1.04996×10^{-10}	4 1.04096×10^{-10}	5 1.67167×10^{-10}
0.99	5 1.67717×10^{-8}	5 1.68163×10^{-8}	5 1.72618×10^{-8}	5 2.17168×10^{-8}
0.9	6 3.50764×10^{-6}	6 3.55624×10^{-6}	6 3.66622×10^{-6}	7 4.23321×10^{-6}
0.8	8 2.13117×10^{-5}	8 2.13854×10^{-5}	8 2.21227×10^{-5}	9 2.64541×10^{-5}
0.7	10 7.41289×10^{-5}	10 7.42528×10^{-5}	10 7.54927×10^{-5}	10 8.78914×10^{-5}

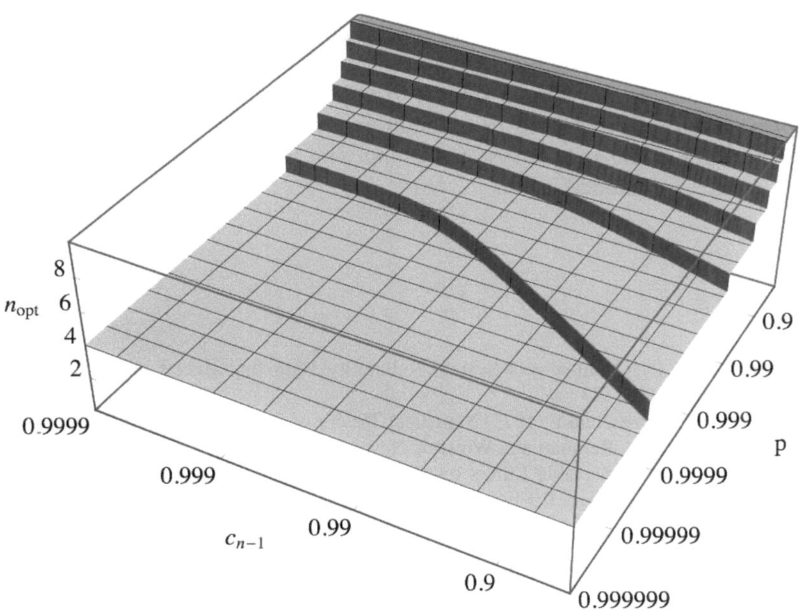

Fig. 10.2 FLC 1-out-of-n:G optimum redundancy level, n_{opt}

10.4 Comparison of Optimum ELC and FLC Systems

Clearly, FLC systems have a greater capacity than ELC systems to exploit increased levels of redundancy for the enhancement of system reliability. For relatively high levels of component reliability ($p > 0.99$ and BIT coverage $c_{n-1} \leq 0.9999$), n_{opt} for an FLC 1-out-of-n:G system is either four or five. Additionally, for a given value of redundant component reliability, an FLC system provides far better reliability than an ELC system.

These results show that if the system designer requires very high levels of system reliability, an FLC design can be used in lieu of an ELC design to gain the benefit of increased component redundancy and fault coverage. This, of course, assumes that such a design choice is possible; such an assumption may not always be the case. The use of an FLC design requires that the system RM be able to effectively "vote" the redundant components. The optimum redundancy level, n_{opt}, for an FLC system is shown in Figure 10.2 as a function of both coverage and component reliability. It can also be seen that the only way to achieve high levels of reliability in an ELC system is to concentrate on component reliability, since an increasing redundancy level quickly reaches a limit in its ability to enhance system reliability.

In summary, for a wide range of realistic coverage values and relatively high levels of component reliability, n_{opt} is four or five for FLC systems and two or three for ELC systems. If a system is subject to imperfect fault coverage, its reliability *cannot* be arbitrarily increased using additional redundancy. Additional treatment of optimum IFC k-out-of-n:G systems, along with the determination of their mean time to failure, is included in [3].

References

1. Myers AF (2007) k-out-of-n:G System Reliability With Imperfect Fault Coverage. IEEE Trans Relia 56:464–473
2. Barlow RE and Proschan F (1975) Statistical Theory of Reliability and Life Testing: Probability Models. Holt, Reinhart and Winston, New York
3. Myers A (2008) Achievable Limits on the Reliability of k-out-of-n:G Systems Subject to Imperfect Fault Coverage. IEEE Trans Relia 57:349–354

Chapter 11
General Architectural Considerations

It is always better to pick a road that goes where you want to get to.

Abstract This chapter outlines general architectural guidelines for the design of highly reliable multichannel systems. The rationale for these guidelines are quantitatively assessed using the BDD-based *FCASE* code.

11.1 Background

There are several general rules for system architecture design that, if followed, provide a reasonable foundation for building a highly reliable system. The system designer must always keep in mind that the only reason for redundancy in a system design is meeting the required level of system reliability. This chapter provides a set of general guidelines for selecting the required level of system reliability and for selecting the level of redundancy that provides the desired probability of system failure. The discussion in Chapter 10 made it clear that the optimum redundancy level is almost always either triplex or quadruplex for highly reliable FLC systems. Once the redundancy level has been selected, one of the most important system design imperatives is maintenance of this level of redundancy throughout the design, unless there are no reasonable alternatives. When the redundancy must be reduced in such a case, the components with the reduced redundancy level must be far more reliable than the components present at the baseline redundancy level.

The adverse consequences of failing to maintain the redundancy level are manifold. The most complex and problematic concerns of system redundancy management design invariably include attempts to accommodate changes in redundancy level. In some cases, there is no practical alternative to a reduction in redundancy; a typical example is the actuation system. In general, the maintenance of a redundancy level has a greater effect on system reliability than does the precise failure rate of any of the redundant components that compose the overall system. This chapter uses the generic quadruplex DFBW system described in Chapter 9 as a baseline example in the demonstration of some basic elements of quality system architecture design. The system depicted in Figures 9.1 and 9.2 was shown to meet stringent probability of failure requirements and is an example of a system with good design "balance"; that is, the system's reliability is not dominated by the unreliability of any single component type. Consequently, this system provides a sound basis for quantifying the effect of modifications to the baseline system architecture.

11.2 Redundancy Level

The quadruplex system of Chapter 9 already includes an example of a change in redundancy level: although the portion of the system that is upstream from the actuation system is quadruplex, each control surface includes only two actuators and is therefore duplex. Since the duplex actuators are far more reliable (λ_{ACT} = 5 fpmh) than the typical quad components, the system is able to retain its overall quad nature. This maintenance of quadruplex system character is demonstrated by the equivalent redundancy level (ERL), which is approximately three. Such an ERL value is to be expected for a quad FLC system.[1]

11.2.1 Variations in Actuator Redundancy

The actuators in the baseline quad system analyzed in Section 9.2 are rather reliable (λ_{ACT} = 5 fpmh). Assume that an alternative actuator design with a significantly improved failure rate is being considered (λ_{ACT} = 0.5 fpmh). With the tenfold improvement in actuator reliability, one could argue that the system can be redesigned to use a single actuator per surface. The effect of this change is illustrated in Figure 11.1. The results of this analysis demonstrate that even though the duplex actuators have been replaced with simplex actuators that are an order of magnitude more reliable than those used in the baseline approach, the adverse effect on the probability of system failure is dramatic. PLOC(10) increases from 1.7×10^{-7} to 4×10^{-5}—a 235-fold increase. The slope of the curve for the system with a single actuator per surface now has the character of a simplex system with an ERL of approximately one. Clearly, if this change were to be adopted, it would make little sense to use quad redundancy for the balance of the system, and it would be unlikely that the reliability of the other components could be increased sufficiently to meet a PLOC(10) $\leq 5 \times 10^{-7}$ design requirement.

This modified simplex actuator system offers a good example of the circumstances in which a system could have the nominal appearance of a quadruplex system while in actuality having the reliability performance of a simplex system.

---

[1] The effect of the duplex actuators can be seen in the change of ERL from 2.24 for a mission time of 1 hour to 2.89 for a mission time of 10 hours. If the actuators were somewhat less reliable, this reduction in ERL would be even more pronounced for low mission times and would be evident at higher mission times as well.

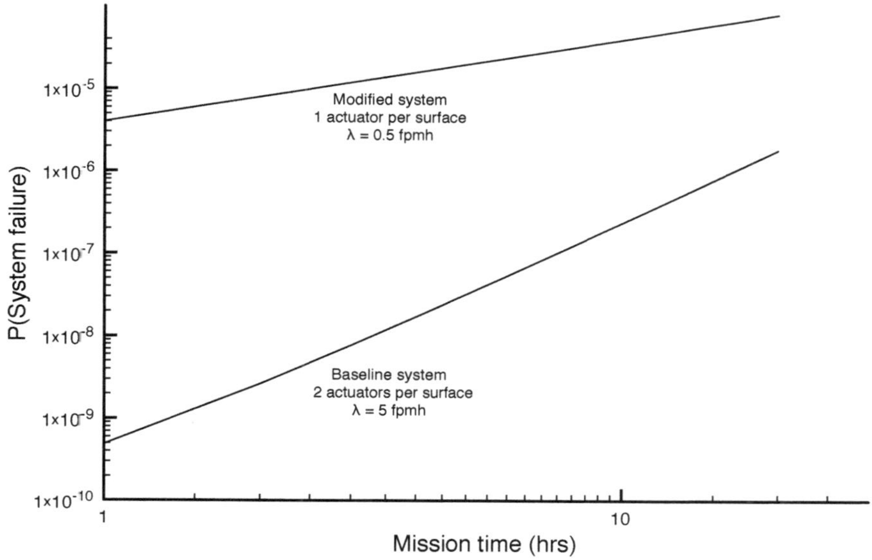

Fig. 11.1 Quad digital flight control systems with one and two actuators per surface

11.2.2 Variations in Hydraulic System Redundancy

The baseline system uses four independent hydraulic pumps, and the overall system is plumbed so that it will continue to operate even if only a single hydraulic system is functioning. One might reason that since the actuators are only dual redundant, two of the four hydraulic systems can be eliminated to make this portion of the system duplex as well. The duplex hydraulic alternative uses two hydraulic pumps, one powered by Engine 1 (ENG1) and the other powered by Engine 3 (ENG3); otherwise, this alternative is identical to the baseline quad system. The probability of system failure for the alternative system and for the baseline system is shown in Figure 11.2. Analysis of the alternative design with only two hydraulic systems shows that the overall system now has the characteristics of a duplex system with an ERL of approximately two.

Figures 11.3 and 11.4 show the *FCASE* results for the contribution to unreliability of the (baseline) quadruplex and (alternative) duplex hydraulic systems. Comparison of each component type's contribution to system unreliability for these two alternative systems shows that the engines (ENG) and pump components (HYD) now dominate the system unreliability, unlike their relatively balanced contribution that characterizes the baseline system. Since each pump is driven by an engine, reducing the number of pumps from four to two dramatically increases the effect of engine reliability on overall system reliability: now one out of two engines, instead

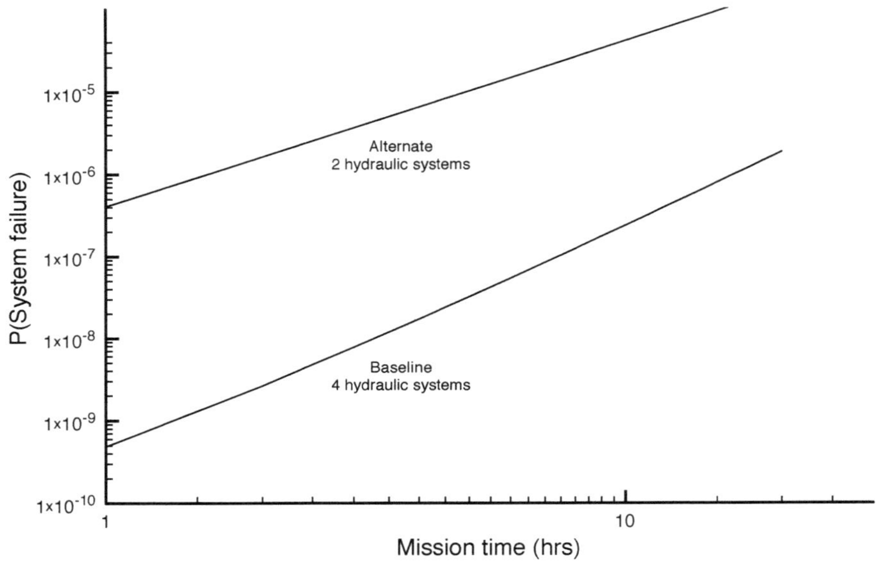

Fig. 11.2 Quad digital flight control systems with two and four hydraulic systems

```
Contribution to Unreliability (sSys) at t = 1.0
        Baseline Q = 3.9843e-10

Variable  Q w/ lambda = 0  Ratio to Baseline

lENG         3.9191e-10         1.0166
lGEN         3.9829e-10         1.0003
lHYD         3.8299e-10         1.0403
lPi          3.5324e-10         1.1279
lPr          3.6682e-10         1.0862
lRIO         3.3679e-10         1.1830
lINS         3.7890e-10         1.0515
lFCC         2.6365e-10         1.5112
lAC          3.8085e-10         1.0461
lSEL         3.9843e-10         1.0000
lACT         1.8224e-10         2.1863
```

Fig. 11.3 *FCASE* results for the contribution to unreliability of the baseline quadruplex hydraulic system

```
Contribution to Unreliability (sSys) at t = 1.0
        Baseline Q = 2.0279e-07

 Variable  Q w/ lambda = 0  Ratio to Baseline

 1ENG          1.2284e-07          1.6509
 1GEN          2.0279e-07          1.0000
 1HYD          1.0381e-08         19.5345
 1Pi           2.0275e-07          1.0002
 1Pr           2.0276e-07          1.0002
 1RIO          2.0273e-07          1.0003
 1INS          2.0277e-07          1.0001
 1FCC          2.0266e-07          1.0007
 1AC           2.0277e-07          1.0001
 1SEL          2.0279e-07          1.0000
 1ACT          2.0259e-07          1.0010
```

Fig. 11.4 *FCASE* results for the contribution to unreliability of the duplex hydraulic system

of one out of four engines, must be operational to avoid system failure. Even if the pumps were perfect (that is, $\lambda_{HYD} = 0$), this system would still have a probability of failure two orders of magnitude greater than that of the baseline system. This analysis provides an excellent example of the critical importance of architecture in the design of highly reliable systems.

11.3 Variations in Redundancy Level

The two alternatives discussed above provide a context for analyzing the effect of changing the redundancy level of two system component types: actuators and hydraulic pumps. Both of these system examples demonstrate the dramatic effect that a change in redundancy level has on the probability of system failure. These results clearly demonstrate the critical importance of maintaining a constant redundancy level throughout the overall system architecture.

The previous examples have demonstrated the difficulty of maintaining a high level of reliability in a system with various levels of redundancy. Obviously, for some level of component reliability, maintaining a high level of system reliability even with varying levels of redundancy should be possible. Consider a system, as shown in Figure 11.5, with four different levels of redundancy. Determining the level of component reliability that is required to evenly distribute the overall system reliability over the A, B, C and D components is illustrative. If each of the component types is i.i.d. and the A components have a failure rate $\lambda_A = 5000$ fpmh, then the A section of this system has a reliability of $1 - 5.65759 \times 10^{-6}$ at $t = 10$ hours. This value can be computed using Equation (2.8). The values for λ_B, λ_C and λ_D that are required to ensure that the B, C and D sections have the same reliability as the

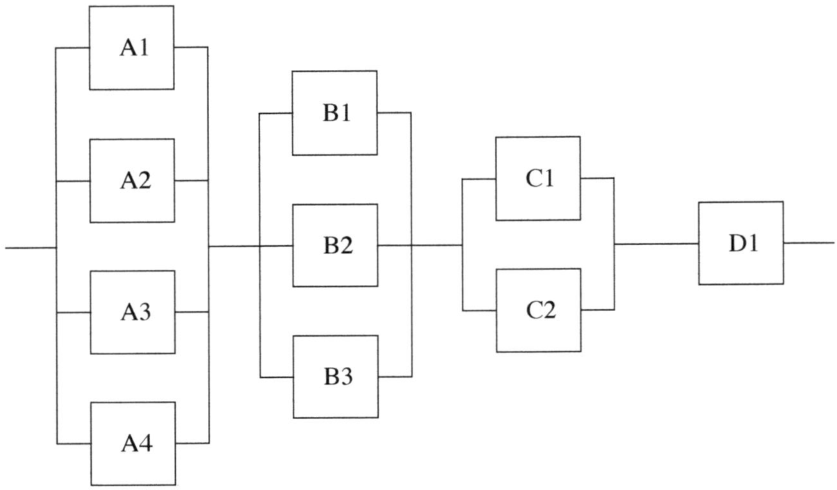

Fig. 11.5 System with varying redundancy levels

A section can also be computed using Equation (2.8). The following equations can then be solved for λ_B, λ_C and λ_D:

$$5.65759 \times 10^{-6} = 1 - e^{-3\lambda_B \cdot 10} + 3e^{-2\lambda_B \cdot 10} - 3e^{-\lambda_B \cdot 10} \quad (11.1)$$

$$5.65759 \times 10^{-6} = 1 + e^{-2\lambda_C \cdot 10} - 2e^{-\lambda_C \cdot 10} \quad (11.2)$$

$$5.65759 \times 10^{-6} = 1 - e^{-\lambda_D \cdot 10} . \quad (11.3)$$

Solving these equations yields $\lambda_B = 1797.941$ fpmh, $\lambda_C = 238.140$ fpmh and $\lambda_D = 0.5657607$ fpmh.

Obviously, the system depicted in Figure 11.5 has a probability of failure of $q_{ABCD} = 1 - (1 - 5.65759 \times 10^{-6})^4 = 2.26302 \times 10^{-5}$ at $t = 10$ hours. The probability of system failure for the A, B, C and D subsystems, along with that of the overall system ABCD, is shown in Figure 11.6. The curves for each of the subsystem unreliabilities intersect at $t = 10$ hours if the λ values computed above are used. The A, B, C and D curves have slopes of 4, 3, 2 and 1 dpd, as expected for quadruplex, triplex, duplex and simplex PFC systems. The overall system, ABCD, asymptotically approaches the 1-out-of-4 subsystem (A) curve for $t > 10$ hours, and it approaches the 1-out-of-1 subsystem (D) curve for $t < 10$ hours. The figure demonstrates that the presence of the 1-out-of-3 subsystem (B) and the 1-out-of-2 subsystem (C) has very little effect on the reliability of the full ABCD system. For $t < 10$ hours, the character of the system is completely dominated by the presence of the "single-point-failure" subsystem, D, and it approaches a slope of 1 dpd. The

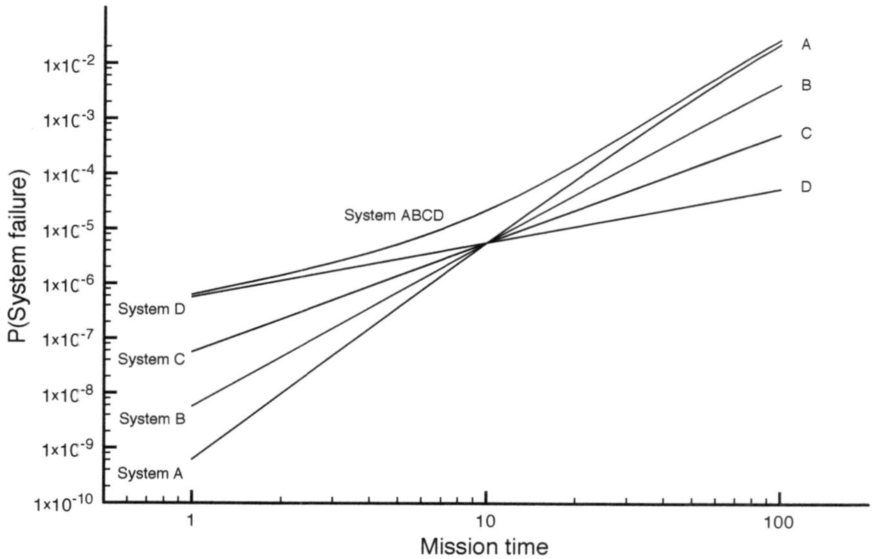

Fig. 11.6 Probability of failure for varying redundancy levels

system takes on the character of a quadruplex system for $t > 10$ hours only when the reliability of the quadruplex (A) system is less than that of the simplex (D) system.

If this experiment were repeated to match the probability of failure for the subsystems at $t = 1$ hour instead of $t = 10$ hours, the required value for λ_D would be 0.000618784 fpmh, and the resulting system would be a highly reliable system. This value for λ_D, however, demonstrates a compelling need to avoid single-point failures in the architectural design of highly reliable systems. It is doubtful that any active component could meet the 0.000618784 fpmh failure rate that is required to match the assigned reliability requirement for component D; in fact, it would require a very robustly designed mechanical part to meet this requirement.[2]

In addition to these challenges posed from a reliability perspective, a reduction in the level of redundancy between two sections creates a difficulty in designing an effective interface between the sections. Consider the interface between sections A and B of the system shown in Figure 11.5, for example. Effective techniques for implementing this interface either require providing the value of all A outputs to each of the B components, then letting the B components vote the A outputs, or they require voting the A components via a CCDL, then sending the voted A value to each of the B components. Both of these alternatives require additional hardware just to support the change in redundancy level. The design of an effective interface and the

[2] The mechanical part would not only have to be robust from a strength perspective, but it would also need to be designed to be virtually immune to fatigue failure as well.

design of the RM implementation in these situations (that is, situations involving changes in redundancy level) are invariably problematic.

11.4 The Value of Cross-Strapping Power

The power distribution system for the quad fly-by-wire system analyzed in Chapter 9 uses four electrical buses (BUS), which are each supplied power from two separate electrical generation units (GEN). This portion of the system is shown in Figure 11.7. In this arrangement, the buses are said to be *cross-strapped.*

Obviously, if the buses can receive power from two independent sources, then the probability that the corresponding channel is powered is higher than it would be in the case of bus power being supplied by a single source. The degree to which this arrangement benefits the overall system reliability may not be clear, however. The architectural decision of whether the electrical buses should be cross-strapped is both important and challenging, since there is a potentially significant penalty in volume, weight and cost. If the system design employs cross-strapping, then the power-generation capacity of each GEN is nominally double the capacity required for the alternative design shown in Figure 11.8, which does not use cross-strapping. In addition to the added cost and weight associated with the cross-strapped design, the designer must also ensure that an electrical fault in a single GEN will not propagate across two buses.

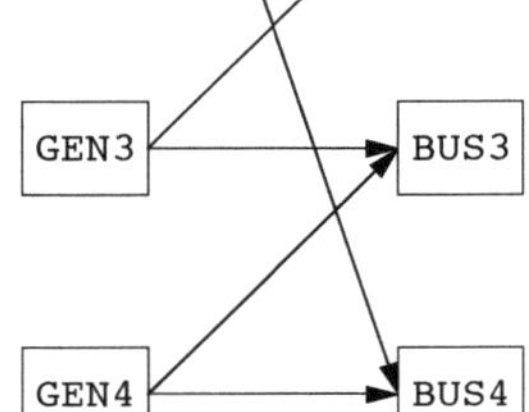

Fig. 11.7 Quad power bus system with cross-strapping

Fortunately, *FCASE* offers a straightforward approach to quantifying the overall effect that these two architectural design alternatives have on system reliability. Re-

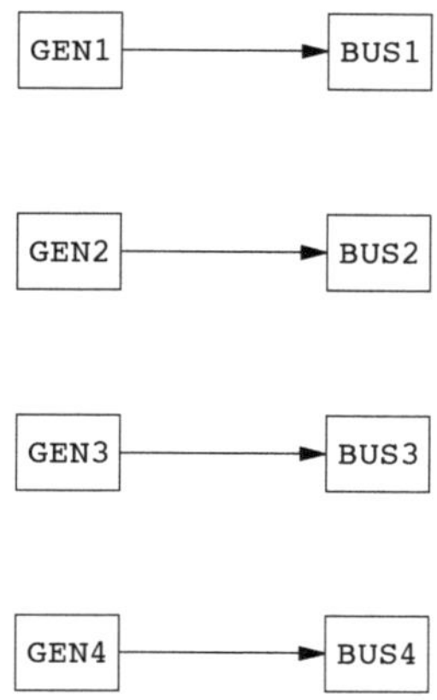

Fig. 11.8 Quad power bus system without cross-strapping

call the following *FCASE* code, which is used to model the cross-strapping of the power buses as shown in Figure 11.7:

```
/* Bus cross-strapping */

sBUS = {
sENGGEN1 | sENGGEN2,
sENGGEN2 | sENGGEN3,
sENGGEN3 | sENGGEN4,
sENGGEN4 | sENGGEN1};
```

The only modification to the `start System` section that is required to model the configuration shown in Figure 11.8 is substitution of the code below:

```
/* BUS is NOT cross-strapped */

sBUS = sENGGEN;
```

If the substituted GEN components have the same failure rate as those that were replaced, then this is the only necessary modification to the *FCASE* code. If the reliability is different, a modification to the `var Def` section is also required.

The probability of overall system failure for the FLC version of the quad fly-by-wire system analyzed in Chapter 9 is shown in Figure 11.9 for both design alternatives. Numerical values are given in Table 11.1; these results were computed under the assumption that the failure rates for the GEN components are the same ($\lambda_{\text{GEN}} = 200$ fpmh).

Eliminating the cross-strapping of the power buses has a significant adverse effect on the probability of system failure: the system without cross-strapping is over three times more likely to fail than is the alternative system with cross-strapping.

Table 11.1 Probability of failure of FLC quad fly-by-wire system with and without power cross-strapping

t (hrs)	With cross-strapping	Without cross-strapping
1	3.98427180^{-10}	9.73973902^{-10}
2	2.08227069^{-9}	6.45214848^{-9}
5	2.37708434^{-8}	9.09328641^{-8}
10	1.69658050^{-7}	7.04408545^{-7}
20	1.29035014^{-6}	5.54296587^{-6}

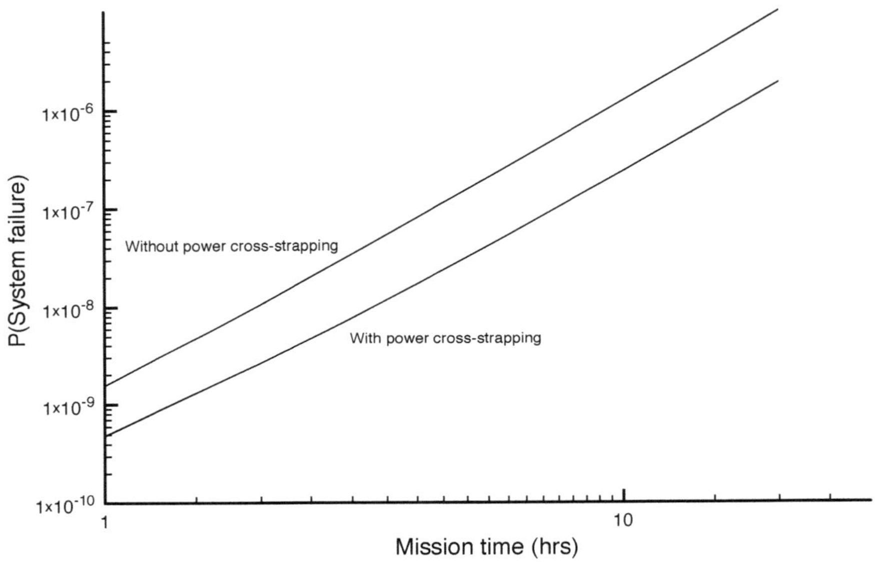

Fig. 11.9 FLC quad fly-by-wire system with and without power cross-strapping

Significantly, the cross-strapped system comfortably meets the design requirement of PLOC(10) $\leq 5 \times 10^{-7}$, but the system without cross-strapping fails to meet this requirement, having PLOC(10) $= 7.04 \times 10^{-7}$. Bus reliability has a significant effect on overall system reliability, since the loss of an electrical bus renders an entire channel inoperative.

11.5 Component Reliability Uncertainty

The architecture of a system has a major effect on its probability of failure. The reliability of a system, of course, is also determined by the reliability of the individual components that make up the overall system. Early in the design process, component failure rates typically must be estimated from whatever data is available; therefore, these estimates may be subject to considerable uncertainty. In general, system architecture (in particular, the uniformity of the redundancy level) has a more significant effect on the resulting overall system reliability than does the precise value of any of the system component failure rates.

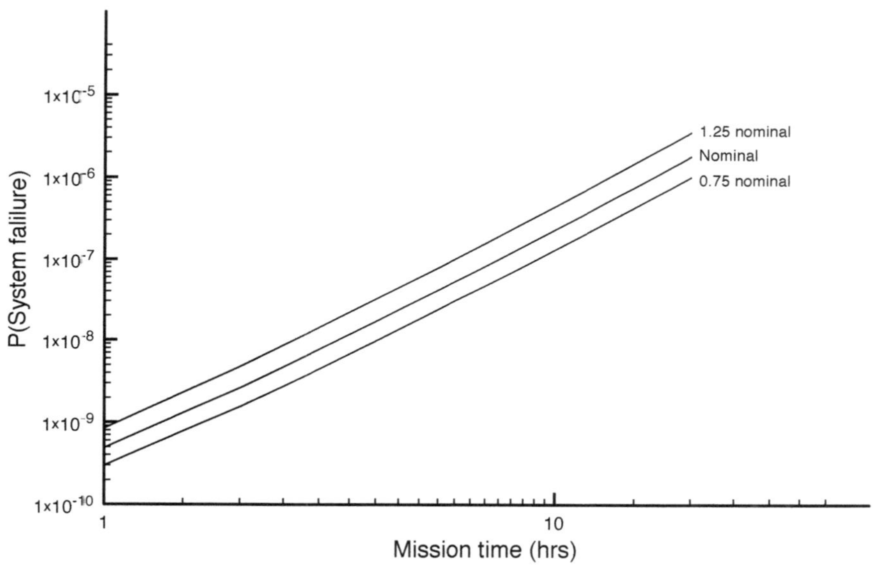

Fig. 11.10 Probability of failure for off-nominal component failure rates

The quad system of Chapter 9 can be used as a baseline to help quantify the effect of component reliability on overall probability of system failure. Figure 11.10 illustrates the effect of the component failure rate on system reliability; three curves are shown: the nominal case, a case with all component failure rates increased to 125% of nominal and a case with all component failure rates reduced to 75% of nominal.

Figure 11.10 demonstrates that a change in component reliability has no significant effect on the character of the probability of system failure: all three curves have nearly the same slope or ERL. By making estimates on the basis of "off-the-shelf" component data and existing system experience, a conservative estimate of the aggregate component reliability within the ±25% range shown in Figure 11.10 should be possible. These results again confirm that the concerns of system architecture are of greater significance to overall system probability of failure than are the precise component failure rates.

Appendix A
Mathematica Combinatorial k-out-of-n:G Functions

The *Mathematica* implementation of both PFC and IFC combinatorial k-out-of-n:G functions are included in this appendix. Even though each of the k-out-of-n:G functions are named **RaL** (at least k-out-of-n:G functions) and **ReX** (exactly k-out-of-n:G functions), *Mathematica* is able to determine the appropriate model (PFC, ELC, FLC or OLC) by examining the arguments.

The combinatorial functions defined here use the **BernoulliRule** substitution, which is applied to the resulting fully expanded polynomial as defined in Section 3.3. These combinatorial functions return correct k-out-of-n:G results even if the vector elements p_i of redundant inputs $\mathbf{p} = \{p_1, \ldots, p_n\}$ are general reliability polynomials are not necessarily disjoint.

A.1 Combinatorial *k*-out-of-*n*:G PFC Functions

```
pT[i_Integer, p_List] := Module[{},
   Apply[Times, KSubsets[p, i], {1}]
  ];

qT[i_Integer, p_List] := Module[{j, n = Length[p], q},
   q = Table[1 - p[[j]], {j, n}];
   Apply[Times, KSubsets[q, n - i], {1}]
  ];

RaL[k_Integer, p_List] := Module[{i, j, n = Length[p]},
   ((Sum[qT[i, p][[j]] pT[i, p][[Binomial[n, i] - j + 1]], {i, k, n}, {j, 1, Binomial[n, i]}])
      // Expand) /. BernoulliRule
  ];

ReX[k_Integer, p_List] := Module[{j, n = Length[p]},
   ((Sum[qT[k, p][[j]] pT[k, p][[Binomial[n, k] - j + 1]], {j, 1, Binomial[n, k]}])
      // Expand) /. BernoulliRule
  ];
```

Fig. A.1 Combinatorial *k*-out-of-*n*:G PFC functions as defined in Equations (2.9) and (2.11)

A.2 Combinatorial *k*-out-of-*n*:G ELC Functions

```
pT[i_Integer, p_List] := Module[{},
   Apply[Times, KSubsets[p, i], {1}]
  ];

qT[i_Integer, p_List] := Module[{j, n = Length[p], q},
   q = Table[1 - p[[j]], {j, n}];
   Apply[Times, KSubsets[q, n - i], {1}]
  ];

cT[i_Integer, c_List] := Module[{j, n = Length[c], cc},
   cc = Table[c[[j]], {j, n}];
   Apply[Times, KSubsets[cc, n - i], {1}]
  ];

RaL[k_Integer, p_List, c_List] := Module[{i, j, n = Length[p]},
  ((Sum[Sum[cT[i, c][[j]] qT[i, p][[j]] pT[i, p][[Binomial[n, i] - j + 1]], {j, 1, Binomial[n, i]}], {i, k, n}])
     // Expand) /. BernoulliRule
 ]; Length[p] == Length[c];

ReX[k_Integer, p_List, c_List] := Module[{j, n = Length[p]},
  ((Sum[cT[k, c][[j]] qT[k, p][[j]] pT[k, p][[Binomial[n, k] - j + 1]], {j, 1, Binomial[n, k]}])
     // Expand) /. BernoulliRule
 ]; Length[p] == Length[c];
```

Fig. A.2 Combinatorial k-out-of-n:G ELC functions as defined in Equations (4.3) and (4.4)

A.3 Combinatorial *k*-out-of-*n*:G FLC Functions

```
pT[i_Integer, p_List] := Module[{},
   Apply[Times, KSubsets[p, i], {1}]
  ];

qT[i_Integer, p_List] := Module[{j, n = Length[p], q},
   q = Table[1 - p[[j]], {j, n}];
   Apply[Times, KSubsets[q, n - i], {1}]
  ];

cP[k_: Integer, n_: Integer, c_: List] := Module[{i},
   Product[c[[i]], {i, 1, n - k}]
  ];

RaL[k_Integer, p_List, c_List] := Module[{i, j, n = Length[p]},
  ((Sum[cP[i, n, c] Sum[qT[i, p][[j]] pT[i, p][[Binomial[n, i] - j + 1]], {j, 1, Binomial[n, i]}], {i, k, n}])
     // Expand) /. BernoulliRule
 ]; Length[p] == (Length[c] + 1);

ReX[k_Integer, p_List, c_List] := Module[{j, n = Length[p]},
  ((cP[k, n, c] Sum[qT[k, p][[j]] pT[k, p][[Binomial[n, k] - j + 1]], {j, 1, Binomial[n, k]}])
     // Expand) /. BernoulliRule
 ]; Length[p] == (Length[c] + 1);
```

Fig. A.3 Combinatorial *k*-out-of-*n*:G FLC functions as defined in Equations (4.8) and (4.9)

A.4 Combinatorial *k*-out-of-*n*:G OLC Functions

```
pT[i_Integer, p_List] := Module[{},
   Apply[Times, KSubsets[p, i], {1}]
  ];

qT[i_Integer, p_List] := Module[{j, n = Length[p], q},
   q = Table[1 - p[[j]], {j, n}];
   Apply[Times, KSubsets[q, n - i], {1}]
  ];

CC[i_: Integer, c_] := Module[{},
   If[i == 1, c, 1]
  ];

RaL[k_Integer, p_List, c_] := Module[{i, j, n = Length[p]},
     (( Sum[CC[i, c] Sum[qT[i, p][[j]] pT[i, p][[Binomial[n,i]-j+1]], {j, 1, Binomial[n,i]}], {i, k, n}] )
        // Expand) /. BernoulliRule
   ] /; AtomQ[c];

ReX[k_Integer, p_List, c_] := Module[{j, n = Length[p]},
     (( CC[k, c] Sum[qT[k, p][[j]] pT[k, p][[Binomial[n,k]-j+1]], {j, 1, Binomial[n,k]}] )
        // Expand) /. BernoulliRule
   ] /; AtomQ[c];
```

Fig. A.4 Combinatorial *k*-out-of-*n*:G OLC functions as defined in Equations (4.13) and (4.14)

Appendix B
Mathematica Recursive k-out-of-n:G Functions

The *Mathematica* implementations of both PFC and IFC recursive k-out-of-n:G functions are included in this appendix. Note that the recursive functions given here assume that the vector **p** of component reliabilities and the vector **c** of coverage values are both defined globally relative to the functions.

The recursive functions defined here use the $\otimes$ and $\oplus$ operators, as defined in Section 3.4, and return correct k-out-of-n:G results even if the vector elements p_i of redundant inputs $\mathbf{p} = \{p_1, \ldots, p_n\}$ are general reliability polynomials that are not necessarrily disjoint.

B.1 Recursive *k*-out-of-*n*:G PFC Functions

```
Rpfc[k_: Integer, n_: Integer] := Module[{},

    If[(n == 0) ∧ (k > n), Return[0]];
    If[(n == 0) ∧ (k ≤ n), Return[1]];

    (1 - p[[n]]) ⊗ Rpfc[k, n - 1] ⊕ p[[n]] ⊗ Rpfc[k - 1, n - 1]
  ];

REpfc[k_: Integer, n_: Integer] := Module[{},

    If[(n < k) ∨ (k < 0), Return[0]];
    If[(n == 0) ∧ (k == n), Return[1]];

    (1 - p[[n]]) ⊗ REpfc[k, n - 1] ⊕ p[[n]] ⊗ REpfc[k - 1, n - 1]
  ];
```

Fig. B.1 Recursive *k*-out-of-*n*:G PFC functions as defined in Section 4.6.1

The vector **p**, which has length **n**, is defined globally outside of the **Rpfc** and **REpfc** function routines.

B.2 Recursive k-out-of-n:G ELC Functions

```
Relc[k_: Integer, n_: Integer] := Module[{},

    If[(n == 0) ⋀ (k > n), Return[0]];
    If[(n == 0) ⋀ (k ≤ n), Return[1]];

    (1 - p[[n]]) ⊗ c[[n]] ⊗ Relc[k, n - 1] ⊕ p[[n]] ⊗ Relc[k - 1, n - 1]
  ];

REelc[k_: Integer, n_: Integer] := Module[{},

    If[(n < k) ⋁ (k < 0), Return[0]];
    If[(n == 0) ⋀ (k == n), Return[1]];

    (1 - p[[n]]) ⊗ c[[n]] ⊗ REelc[k, n - 1] ⊕ p[[n]] ⊗ REelc[k - 1, n - 1]
  ];
```

Fig. B.2 Recursive k-out-of-n:G ELC functions as defined in Section 4.6.2

The vectors **p** and **c**, both of length **n**, are defined globally outside of the **Relc** and **REelc** function routines. The vector $\mathbf{c} = \{c_1, \ldots, c_n\}$ contains the ELC coverage values.

B.3 Recursive *k*-out-of-*n*:G FLC Functions

```
Rflc[k_: Integer, n_: Integer, f_: Integer] := Module[{},

    If[(n == 0) ∧ (k > n), Return[0]];
    If[(n == 0) ∧ (k ≤ n), Return[1]];

    (1 - p[[n]]) ⊗ c[[f+1]] ⊗ Rflc[k, n - 1, f + 1] ⊕ p[[n]] ⊗ Rflc[k - 1, n - 1, f]
  ];

REflc[k_: Integer, n_: Integer, f_: Integer] := Module[{},

    If[(n < k) ∨ (k < 0), Return[0]];
    If[(n == 0) ∧ (k == n), Return[1]];

    (1 - p[[n]]) ⊗ c[[f+1]] ⊗ REflc[k, n - 1, f + 1] ⊕ p[[n]] ⊗ REflc[k - 1, n - 1, f]
  ];
```

Fig. B.3 Recursive *k*-out-of-*n*:G FLC functions as defined in Section 4.6.3

The vectors **p** and **c,** both of length **n,** are defined globally outside of the **Rflc** and **REflc** function routines. The vector $\mathbf{c} = \{c_1, \ldots, c_{n-1}, 0\}$ contains the FLC coverage values. The functions **Rflc** and **REflc** are called with $f = 0$ to obtain FLC *k*-out-of-*n*:G reliabilities.

B.4 Recursive *k*-out-of-*n*:G OLC Functions

```
Rolc[k_: Integer, n_: Integer, f_: Integer] := Module[{cov},

    If[(n == 0) ∧ (k > n), Return[0]];
    If[(n == 0) ∧ (k ≤ n), Return[1]];

    cov = If[(f + 1) == (Length[p] - 1), c, 1];

    (1 - p[[n]]) ⊗ cov ⊗ Rolc[k, n - 1, f + 1] ⊕ p[[n]] ⊗ Rolc[k - 1, n - 1, f]
  ];

REolc[k_: Integer, n_: Integer, f_: Integer] := Module[{cov},

    If[(n < k) ∨ (k < 0), Return[0]];
    If[(n == 0) ∧ (k == n), Return[1]];

    cov = If[(f + 1) == (Length[p] - 1), c, 1];

    (1 - p[[n]]) ⊗ cov ⊗ REolc[k, n - 1, f + 1] ⊕ p[[n]] ⊗ REolc[k - 1, n - 1, f]
  ];
```

Fig. B.4 Recursive k-out-of-n:G OLC functions as defined in Section 4.6.4

The vector **p**, which has length `n`, and the scalar c are defined globally outside of the `Rolc` and `REolc` function routines. The scalar c is the OLC coverage value. The functions `Rolc` and `REolc` are called with $f = 0$ to obtain OLC k-out-of-n:G reliabilities.

Appendix C
Mathematica Table-Based k-out-of-n:G Functions

The *Mathematica* implementation of both PFC and IFC table-based k-out-of-n:G functions are included in this appendix. Even though each of the k-out-of-n:G functions are named **RaL** (at least k-out-of-n:G functions) and **ReX** (exactly k-out-of-n:G functions), *Mathematica* is able to determine the appropriate model (PFC, ELC, FLC or OLC) by examining the arguments.

The table-based routines given here use the $\otimes$ and $\oplus$ operators, as defined in Section 3.4, and return correct k-out-of-n:G results even if the vector elements p_i of redundant inputs $\mathbf{p} = \{p_1, \ldots, p_n\}$ are general reliability polynomials that are not necessarily disjoint.

C.1 Table-Based *k*-out-of-*n*:G PFC Functions

```
RaL[k_, p_List] := Module[{P, r, i, j, n},

   n = Length[p];
   P = Table[0, {n - k + 1}];
   P[[1]] = 1;
   r = 0;

   For[i = 1, i ≤ n, i++,
    For[j = n - k, j ≥ 1, j --,
     P[[j+1]] = p[[i]] ⊗ P[[j+1]] ⊕ (1 - p[[i]]) ⊗ P[[j]];
     If[i == n, r = r ⊕ P[[j+1]]];
    ];
    P[[1]] = P[[1]] ⊗ p[[i]];
   ];

   r = r ⊕ P[[1]]

  ];

ReX[k_, p_List] := Module[{P, i, j, n},

   n = Length[p];
   P = Table[0, {n - k + 1}];
   P[[1]] = 1;

   For[i = 1, i ≤ n, i++,
    For[j = n - k, j ≥ 1, j --,
     P[[j+1]] = p[[i]] ⊗ P[[j+1]] ⊕ (1 - p[[i]]) ⊗ P[[j]];
     If[i == n, Return[P[[j+1]]]];
    ];
    P[[1]] = P[[1]] ⊗ p[[i]];
   ];

   P[[1]]

  ];
```

Fig. C.1 Table-based *k*-out-of-*n*:G PFC functions as defined in Section 4.7.1

C.2 Table-Based *k*-out-of-*n*:G ELC Functions

```
RaL[k_, p_List, c_List] := Module[{P, r, i, j, n},

    n = Length[p];
    P = Table[0, {n - k + 1}];
    P[[1]] = 1;
    r = 0;

    For[i = 1, i ≤ n, i++,
     For[j = n - k, j ≥ 1, j--,
      P[[j+1]] = p[[i]] ⊗ P[[j+1]] ⊕ c[[i]] ⊗ (1 - p[[i]]) ⊗ P[[j]];
      If[i == n, r = r ⊕ P[[j+1]]];
     ];
     P[[1]] = P[[1]] ⊗ p[[i]];
    ];

    r = r ⊕ P[[1]]

   ] /; Length[p] == Length[c];

ReX[k_, p_List, c_List] := Module[{P, i, j, n},

    n = Length[p];
    P = Table[0, {n - k + 1}];
    P[[1]] = 1;

    For[i = 1, i ≤ n, i++,
     For[j = n - k, j ≥ 1, j--,
      P[[j+1]] = p[[i]] ⊗ P[[j+1]] ⊕ c[[i]] ⊗ (1 - p[[i]]) ⊗ P[[j]];
      If[i == n, Return[P[[j+1]]]];
     ];
     P[[1]] = P[[1]] ⊗ p[[i]];
    ];

    P[[1]]

   ] /; Length[p] == Length[c];
```

Fig. C.2 Table-based *k*-out-of-*n*:G ELC functions as defined in Section 4.7.2

C.3 Table-Based *k*-out-of-*n*:G FLC Functions

```
RaL[k_, p_List, c_List] := Module[{P, r, i, j, n},

    n = Length[p];
    P = Table[0, {n - k + 1}];
    P[[1]] = 1;
    r = 0;

    For[i = 1, i ≤ n, i ++,
     For[j = n - k, j ≥ 1, j --,
      P[[j+1]] = p[[i]] ⊗ P[[j+1]] ⊕ c[[j]] ⊗ (1 - p[[i]]) ⊗ P[[j]];
      If[i == n, r = r ⊕ P[[j+1]]];
     ];
     P[[1]] = P[[1]] ⊗ p[[i]];
    ];

    r = r ⊕ P[[1]]

   ] /; (Length[p] - 1) == Length[c];

ReX[k_, p_List, c_List] := Module[{P, i, j, n},

    n = Length[p];
    P = Table[0, {n - k + 1}];
    P[[1]] = 1;

    For[i = 1, i ≤ n, i ++,
     For[j = n - k, j ≥ 1, j --,
      P[[j+1]] = p[[i]] ⊗ P[[j+1]] ⊕ c[[j]] ⊗ (1 - p[[i]]) ⊗ P[[j]];
      If[i == n, Return[P[[j+1]]]];
     ];
     P[[1]] = P[[1]] ⊗ p[[i]];
    ];

    P[[1]]

   ] /; (Length[p] - 1) == Length[c];
```

Fig. C.3 Table-based *k*-out-of-*n*:G FLC functions as defined in Section 4.7.3

C.4 Table-Based *k*-out-of-*n*:G OLC Functions

```
RaL[k_, p_List, c_] := Module[{P, r, i, j, n},

    n = Length[p];
    P = Table[0, {n - k + 1}];
    P[[1]] = 1;
    r = 0;

    For[i = 1, i ≤ n, i ++,
     For[j = n - k, j ≥ 1, j --,
      If[j == n - 1,
       P[[j+1]] = p[[i]] ⊗ P[[j+1]] ⊕ c ⊗ (1 - p[[i]]) ⊗ P[[j]],
       P[[j+1]] = p[[i]] ⊗ P[[j+1]] ⊕ (1 - p[[i]]) ⊗ P[[j]]];
      If[i == n, r = r ⊕ P[[j+1]]];
     ];
     P[[1]] = P[[1]] ⊗ p[[i]];
    ];

    r = r ⊕ P[[1]]

   ] /; AtomQ[c];

ReX[k_, p_List, c_] := Module[{P, i, j, n},

    n = Length[p];
    P = Table[0, {n - k + 1}];
    P[[1]] = 1;

    For[i = 1, i ≤ n, i ++,
     For[j = n - k, j ≥ 1, j --,
      If[j == n - 1,
       P[[j+1]] = p[[i]] ⊗ P[[j+1]] ⊕ c ⊗ (1 - p[[i]]) ⊗ P[[j]],
       P[[j+1]] = p[[i]] ⊗ P[[j+1]] ⊕ (1 - p[[i]]) ⊗ P[[j]]];
      If[i == n, Return[P[[j+1]]]];
     ];
     P[[1]] = P[[1]] ⊗ p[[i]];
    ];

    P[[1]]

   ] /; AtomQ[c];
```

Fig. C.4 Table-based *k*-out-of-*n*:G OLC functions as defined in Section 4.7.4

Appendix D
FCASE Implementation of System A and System B

Section 8.4 discusses the use of *FCASE* to assess the reliability of System A and System B. This appendix includes the full listing of the *FCASE* files that provide models for these systems. The *FCASE* models shown here correspond to the CPM analyses of System A and System B that were developed in Chapter 6. The numerical data required for the assessment of both System A and System B are the same as those used in Chapter 5, given in Tables 5.1, 5.2 and 5.4.

D.1 *FCASE* System A

System A is shown in Figure 6.2.

```
                                FCASE
Version: 07_17_09

 ************ Flight Critical Aircraft System Evaluation ***
*********


      Run date: Thu Jul 30 09:15:05 2009

      Elapsed time: 0 seconds
Input string to FCASE App


    /* Combined computer system and actuation system */
    /* Book example System A - FLC */

    start VarDef;
```

```
lP = 200e-6;
lS = 250e-6;
lC = 400e-6;

lAC = 400e-6;
lE = 10e-6;
lA = 2e-6;

cSolc = 0.9;
cColc = 0.95;
cAColc = 0.95;
cEolc = 0.9;

cS = {covCal(4, 1, lS, 30),
      covCal(4, 2, lS, 30), cSolc};
cC = {covCal(4, 1, lC, 30),
      covCal(4, 2, lC, 30), cColc};

cAC = {covCal(4, 1, lAC, 30),
       covCal(4, 2, lAC, 30), cAColc};
cEL = {covCal(4, 1, lE, 30),
       covCal(4, 2, lE, 30), cEolc};
cER = {covCal(4, 1, lE, 30),
       covCal(4, 2, lE, 30), cEolc};

bddVarDef(P, 4, exp, lP);
bddVarDef(S, 4, exp, lS);
bddVarDef(cS, cS);
bddVarDef(C, 4, exp, lC);
bddVarDef(cC, cC);

/* Surface A */

bddVarDef(ACa, 4, exp, lAC);
bddVarDef(cACa, cAC);
bddVarDef(ELa, 4, exp, lE);
bddVarDef(cELa, cEL);
bddVarDef(ERa, 4, exp, lE);
bddVarDef(cERa, cER);
bddVarDef(Aa, 2, exp, lA);

/* Surface B */

bddVarDef(ACb, 4, exp, lAC);
```

```
bddVarDef(cACb, cAC);
bddVarDef(ELb, 4, exp, lE);
bddVarDef(cELb, cEL);
bddVarDef(ERb, 4, exp, lE);
bddVarDef(cERb, cER);
bddVarDef(Ab, 2, exp, lA);

/* Surface C */

bddVarDef(ACc, 4, exp, lAC);
bddVarDef(cACc, cAC);
bddVarDef(ELc, 4, exp, lE);
bddVarDef(cELc, cEL);
bddVarDef(ERc, 4, exp, lE);
bddVarDef(cERc, cER);
bddVarDef(Ac, 2, exp, lA);

/* Surface D */

bddVarDef(ACd, 4, exp, lAC);
bddVarDef(cACd, cAC);
bddVarDef(ELd, 4, exp, lE);
bddVarDef(cELd, cEL);
bddVarDef(ERd, 4, exp, lE);
bddVarDef(cERd, cER);
bddVarDef(Ad, 2, exp, lA);

/* Surface E */

bddVarDef(ACe, 4, exp, lAC);
bddVarDef(cACe, cAC);
bddVarDef(ELe, 4, exp, lE);
bddVarDef(cELe, cEL);
bddVarDef(ERe, 4, exp, lE);
bddVarDef(cERe, cER);
bddVarDef(Ae, 2, exp, lA);

/* Surface F */

bddVarDef(ACf, 4, exp, lAC);
bddVarDef(cACf, cAC);
bddVarDef(ELf, 4, exp, lE);
bddVarDef(cELf, cEL);
bddVarDef(ERf, 4, exp, lE);
bddVarDef(cERf, cER);
```

```
bddVarDef(Af, 2, exp, 1A);

start System

/* Combined problem */

sBout = {P1 | P4, P2 | P1, P3 | P2, P4 | P3};

sSpwr = S & sBout;
sCpwr = C & sBout;

sSin = sSpwr & C;

sSvoted = FLC(1, sSin, cS);

sCin = sSvoted & sCpwr;

sCvoted = FLC(1, sCin, cC);

/* Surface A */

sACpwra = ACa & sBout;
sACouta = sCvoted & FLC(1, sACpwra, cACa);
sELouta = sACouta & FLC(1, sACpwra & ELa, cELa);
sERouta = sACouta & FLC(1, sACpwra & ERa, cERa);

sAct1a = Aa1 & sELouta;
sAct2a = Aa2 & sERouta;

sSRFa = sAct1a | sAct2a;

/* Surface B */

sACpwrb = ACb & sBout;
sACoutb = sCvoted & FLC(1, sACpwrb, cACb);
sELoutb = sACoutb & FLC(1, sACpwrb & ELb, cELb);
sERoutb = sACoutb & FLC(1, sACpwrb & ERb, cERb);

sAct1b = Ab1 & sELoutb;
sAct2b = Ab2 & sERoutb;

sSRFb = sAct1b | sAct2b;

/* Surface C */
```

```
sACpwrc = ACc & sBout;
sACoutc = sCvoted & FLC(1, sACpwrc, cACc);
sELoutc = sACoutc & FLC(1, sACpwrc & ELc, cELc);
sERoutc = sACoutc & FLC(1, sACpwrc & ERc, cERc);

sAct1c = Ac1 & sELoutc;
sAct2c = Ac2 & sERoutc;

sSRFc = sAct1c | sAct2c;

/* Surface D */

sACpwrd = ACd & sBout;
sACoutd = sCvoted & FLC(1, sACpwrd, cACd);
sELoutd = sACoutd & FLC(1, sACpwrd & ELd, cELd);
sERoutd = sACoutd & FLC(1, sACpwrd & ERd, cERd);

sAct1d = Ad1 & sELoutd;
sAct2d = Ad2 & sERoutd;

sSRFd = sAct1d | sAct2d;

/* Surface E */

sACpwre = ACe & sBout;
sACoute = sCvoted & FLC(1, sACpwre, cACe);
sELoute = sACoute & FLC(1, sACpwre & ELe, cELe);
sERoute = sACoute & FLC(1, sACpwre & ERe, cERe);

sAct1e = Ae1 & sELoute;
sAct2e = Ae2 & sERoute;

sSRFe = sAct1e | sAct2e;

/* Surface F */

sACpwrf = ACf & sBout;
sACoutf = sCvoted & FLC(1, sACpwrf, cACf);
sELoutf = sACoutf & FLC(1, sACpwrf & ELf, cELf);
sERoutf = sACoutf & FLC(1, sACpwrf & ERf, cERf);

sAct1f = Af1 & sELoutf;
sAct2f = Af2 & sERoutf;

sSRFf = sAct1f | sAct2f;
```

```
sSys = sSRFa & sSRFb & sSRFc & sSRFd & sSRFe & sSRFf;

start Results

repeat(sSRFa, 1, 20, 1);

repeat(sSys, 1, 20, 1);

problem end


    Elapsed time: 0 seconds

Initializing BDD tables

Evaluate BDD node values for basic variables

elapsed time after evaluateBDDbasic: 1 seconds

Evaluating Problem

elapsed time after evaluateProblem: 1 seconds

 Q(sSRFa) for t =  1.00 to  20.00 by incr 1.0000

      1.00    1.88736471e-10
      2.00    1.20550947e-09
      3.00    3.87111410e-09
      4.00    9.00711972e-09
      5.00    1.74358599e-08
      6.00    2.99804220e-08
      7.00    4.74646323e-08
      8.00    7.07130494e-08
      9.00    1.00550950e-07
     10.00    1.37804318e-07
     11.00    1.83299834e-07
     12.00    2.37864866e-07
     13.00    3.02327454e-07
     14.00    3.77516301e-07
```

```
      15.00     4.64260764e-07
      16.00     5.63390840e-07
      17.00     6.75737157e-07
      18.00     8.02130962e-07
      19.00     9.43404109e-07
      20.00     1.10038905e-06

 Q(sSys) for t =  1.00 to  20.00 by incr 1.0000

       1.00     3.64957842e-10
       2.00     2.05605233e-09
       3.00     6.35409203e-09
       4.00     1.45426891e-08
       5.00     2.79082428e-08
       6.00     4.77399272e-08
       7.00     7.53296677e-08
       8.00     1.11972131e-07
       9.00     1.58964704e-07
      10.00     2.17607480e-07
      11.00     2.89203241e-07
      12.00     3.75057441e-07
      13.00     4.76478189e-07
      14.00     5.94776233e-07
      15.00     7.31264945e-07
      16.00     8.87260300e-07
      17.00     1.06408086e-06
      18.00     1.26304777e-06
      19.00     1.48548472e-06
      20.00     1.73271793e-06

                                     FCASE

  ************ Flight Critical Aircraft System Evaluation ***
*********



  ************         Normal End of Evaluation Run       ***
*********

elapsed time at problem end: 7 seconds
```

D.2 *FCASE* System B

System B is shown in Figure 6.3.

```
                                  FCASE
Version: 07_17_09

 ************ Flight Critical Aircraft System Evaluation ***
*********


      Run date: Thu Jul 30 09:21:04 2009

      Elapsed time: 0 seconds
Input string to FCASE App


    /* Combined computer system and actuation system */
    /* Book example System B - FLC */

    start VarDef;

    lP = 200e-6;
    lS = 250e-6;
    lC = 400e-6;

    lAC = 400e-6;
    lE = 10e-6;
    lA = 2e-6;

    cSolc = 0.9;
    cColc = 0.95;
    cAColc = 0.95;
    cEolc = 0.9;

    cS = {covCal(4, 1, lS, 30), covCal(4, 2, lS, 30), cSolc};
    cC = {covCal(4, 1, lC, 30), covCal(4, 2, lC, 30), cColc};

    cAC = {covCal(4, 1, lAC, 30),
           covCal(4, 2, lAC, 30), cAColc};
    cEL = {covCal(4, 1, lE, 30),
           covCal(4, 2, lE, 30), cEolc};
    cER = {covCal(4, 1, lE, 30),
           covCal(4, 2, lE, 30), cEolc};
```

```
bddVarDef(P, 4, exp, lP);
bddVarDef(S, 4, exp, lS);
bddVarDef(cS, cS);
bddVarDef(C, 4, exp, lC);
bddVarDef(cC, cC);
bddVarDef(AC, 4, exp, lAC);
bddVarDef(cAC, cAC);

/* Surface A */

bddVarDef(ELa, 4, exp, lE);
bddVarDef(cELa, cEL);
bddVarDef(ERa, 4, exp, lE);
bddVarDef(cERa, cER);
bddVarDef(Aa, 2, exp, lA);

/* Surface B */

bddVarDef(ELb, 4, exp, lE);
bddVarDef(cELb, cEL);
bddVarDef(ERb, 4, exp, lE);
bddVarDef(cERb, cER);
bddVarDef(Ab, 2, exp, lA);

/* Surface C */

bddVarDef(ELc, 4, exp, lE);
bddVarDef(cELc, cEL);
bddVarDef(ERc, 4, exp, lE);
bddVarDef(cERc, cER);
bddVarDef(Ac, 2, exp, lA);

/* Surface D */

bddVarDef(ELd, 4, exp, lE);
bddVarDef(cELd, cEL);
bddVarDef(ERd, 4, exp, lE);
bddVarDef(cERd, cER);
bddVarDef(Ad, 2, exp, lA);

/* Surface E */

bddVarDef(ELe, 4, exp, lE);
bddVarDef(cELe, cEL);
```

```
bddVarDef(ERe, 4, exp, lE);
bddVarDef(cERe, cER);
bddVarDef(Ae, 2, exp, lA);

/* Surface F */

bddVarDef(ELf, 4, exp, lE);
bddVarDef(cELf, cEL);
bddVarDef(ERf, 4, exp, lE);
bddVarDef(cERf, cER);
bddVarDef(Af, 2, exp, lA);

start System

/* Combined problem */

sBout = {P1 | P4, P2 | P1, P3 | P2, P4 | P3};

sSpwr = S & sBout;
sCpwr = C & sBout;
sACpwr = AC & sBout;

sSin = sSpwr & C;

sSvoted = FLC(1, sSin, cS);

sCin = sSvoted & sCpwr;

sCvoted = FLC(1, sCin, cC);

sACvoted = FLC(1, sCvoted & sACpwr, cAC);

sACout = sACvoted & sACpwr;

/* Surface A */

sELouta = FLC(1, sACout & ELa, cELa);
sERouta = FLC(1, sACout & ERa, cERa);

sAct1a = Aa1 & sELouta;
sAct2a = Aa2 & sERouta;

sSRFa = sAct1a | sAct2a;

/* Surface B */
```

```
sELoutb = FLC(1, sACout & ELb, cELb);
sERoutb = FLC(1, sACout & ERb, cERb);

sAct1b = Ab1 & sELoutb;
sAct2b = Ab2 & sERoutb;

sSRFb = sAct1b | sAct2b;

/* Surface C */

sELoutc = FLC(1, sACout & ELc, cELc);
sERoutc = FLC(1, sACout & ERc, cERc);

sAct1c = Ac1 & sELoutc;
sAct2c = Ac2 & sERoutc;

sSRFc = sAct1c | sAct2c;

/* Surface D */

sELoutd = FLC(1, sACout & ELd, cELd);
sERoutd = FLC(1, sACout & ERd, cERd);

sAct1d = Ad1 & sELoutd;
sAct2d = Ad2 & sERoutd;

sSRFd = sAct1d | sAct2d;

/* Surface E */

sELoute = FLC(1, sACout & ELe, cELe);
sERoute = FLC(1, sACout & ERe, cERe);

sAct1e = Ae1 & sELoute;
sAct2e = Ae2 & sERoute;

sSRFe = sAct1e | sAct2e;

/* Surface F */

sELoutf = FLC(1, sACout & ELf, cELf);
sERoutf = FLC(1, sACout & ERf, cERf);

sAct1f = Af1 & sELoutf;
```

```
sAct2f = Af2 & sERoutf;

sSRFf = sAct1f | sAct2f;

sSys = sSRFa & sSRFb & sSRFc & sSRFd & sSRFe & sSRFf;

start Results

repeat(sSRFa, 1, 20, 1);

repeat(sSys, 1, 20, 1);

problem end
```

```
    Elapsed time: 0 seconds

Initializing BDD tables

Evaluate BDD node values for basic variables

elapsed time after evaluateBDDbasic: 1 seconds

Evaluating Problem

elapsed time after evaluateProblem: 2 seconds

 Q(sSRFa) for t =  1.00 to  20.00 by incr 1.0000

      1.00    1.88736471e-10
      2.00    1.20550947e-09
      3.00    3.87111410e-09
      4.00    9.00711972e-09
      5.00    1.74358599e-08
      6.00    2.99804220e-08
      7.00    4.74646323e-08
      8.00    7.07130494e-08
      9.00    1.00550950e-07
     10.00    1.37804318e-07
```

```
      11.00     1.83299834e-07
      12.00     2.37864866e-07
      13.00     3.02327454e-07
      14.00     3.77516301e-07
      15.00     4.64260764e-07
      16.00     5.63390840e-07
      17.00     6.75737157e-07
      18.00     8.02130962e-07
      19.00     9.43404109e-07
      20.00     1.10038905e-06

 Q(sSys) for t =  1.00 to  20.00 by incr 1.0000

       1.00     2.20532037e-10
       2.00     1.37983680e-09
       3.00     4.36934433e-09
       4.00     1.00811506e-08
       5.00     1.94080040e-08
       6.00     3.32432956e-08
       7.00     5.24810443e-08
       8.00     7.80158906e-08
       9.00     1.10743079e-07
      10.00     1.51558451e-07
      11.00     2.01358429e-07
      12.00     2.61040009e-07
      13.00     3.31500750e-07
      14.00     4.13638755e-07
      15.00     5.08352668e-07
      16.00     6.16541658e-07
      17.00     7.39105407e-07
      18.00     8.76944104e-07
      19.00     1.03095842e-06
      20.00     1.20204953e-06

                                FCASE

  ************ Flight Critical Aircraft System Evaluation ***
 *********



  ************          Normal End of Evaluation Run       ***
 *********

elapsed time at problem end: 7 seconds
```

Appendix E
FCASE Input File Syntax

FCASE is a Unix command-line tool that uses binary decision diagram techniques to develop exact models for system reliability. Input to the *FCASE* program is in the form of an ASCII text file, and its output is ASCII text in the Unix command-line terminal environment. Since *FCASE* is written in ANSI standard C, it can be used on any computer platform that supports the C language in a command-line environment. Additional information regarding *FCASE* can be found at `amyersconsulting.com`.

The *FCASE* input file consists of a sequence of commands that are used to define the variables that describe the system (along with the ordering of these variables for building the BDD), the structure of the system being modeled, the requested output for the results and a termination point that indicates the end of the problem. Each of these sections is defined by a single-line section header occurring in the following order:

- `start VarDef`
- `start System`
- `start Results`
- `problem end`

Although each of the *FCASE* commands contained in these sections is terminated with a semicolon (`;`), the section header lines are entered exactly as shown above, with no terminating character. A description of the *FCASE* commands that are used in each of the sections is given below.

E.1 *FCASE* `start VarDef` Section

The first portion of the `start VarDef` section is used to define the numerical constants that describe the system, such as the component failure rates and the coverage values. These constants are defined in the same manner as they would be defined in C or FORTRAN. For instance, to define a failure rate of $\lambda_p = 1000$ fpmh for a set of i.i.d. components p_i, the following command can be used:

```
lP = 1000e-6;
```

Like all *FCASE* commands, the statement is terminated with a semicolon.

Coverage values are defined in the same manner. For example, if the vector $\boldsymbol{p}$ of quad-redundant components is associated with a vector `cPelc` of ELC coverage values, then `cPelc` can be defined as

```
cPelc = {0.9, 0.9, 0.9, 0.9};
```

Here, the variable `cPelc` is defined as a vector of length four with each of its elements set to a value of 0.9. The coverage vector for an FLC system, `cPflc`, either can be defined numerically using

```
cPflc = {0.999999975, 0.999999983, 0.9};
```

or can be computed with the `covCal` function based on Equation (4.39) using

```
cPflc = {covCal(4, 1, lP, 30), covCal(4, 2, lP, 30), 0.9};
```

The `covCal` arguments are the same as those defined on page 59 for the corresponding *Mathematica* **covCal** function. An OLC coverage value (a scalar) can be defined as

```
cPolc = 0.9;
```

In these cases, the variables `cPelc`, `cPflc` and `cPolc` have been given particular descriptive names. This naming scheme, however, is not a requirement, and the user is free to assign any desired variable name, subject to the conditions that the name is alphanumeric and that it starts with an alpha character.

Once all required numerical values are assigned, the problem definition variables are defined. The variable ordering, which is required for the BDD construction, is determined by the order in which the problem variables are defined.

The *FCASE* function that is used to simultaneously define the problem variables is `bddVarDef`. The sequence in which the `bddVarDef` calls are made determines the variable ordering. Variables, both scalar and vector, that have an exponential reliability function are defined using the `bddVarDef` function. For example, if the vector $\boldsymbol{p}$ represents a quad-redundant set of scalars that is written as $\{p_1, p_2, p_3, p_4\}$, then these scalar values can be defined as

```
bddVarDef(p, 4, exp, lP);
```

Here, the first argument defines the variable name, the second argument defines the length of the vector, the third argument defines the reliability function type and the fourth argument defines the failure rate. If the second argument is unity, then the variable being defined is a scalar.

Some variables, such as coverage values, represent probabilities that are not functions of time. For instance, the ELC coverage vector given as an example above can be defined as

```
bddVarDef(cPelc, cPelc);
```

In this case, the problem variable has been given the same name as that used to define the numerical value `cPelc`. This naming approach is not essential, however; the variable can be assigned any legitimate *FCASE* variable name (again, as long as it is alphanumeric and starts with an alpha character). In this example, *FCASE* knows that the variable being defined, `cPelc`, is a vector of length four because the constant variable, also named `cPelc`, was earlier defined as a vector.

A scalar is defined in the same fashion:

```
bddVarDef(cPolc, cPolc);
```

Again, the variable `cPolc` is known to be a scalar because the associated constant variable was defined as a scalar.

The efficiency, both in terms of size and computational time, is strongly dependent on the ordering of the variables that are used to define the problem. Even though the determination of an optimum variable ordering is known to be an NP-hard problem, experience has demonstrated that the following straightforward convention yields good efficiency.

1. Select the variable ordering as it occurs in a functional block diagram from left to right (that is, from upstream to downstream).
2. Place the coverage variables immediately after the associated component vector.

Using these simple ordering rules, *FCASE* has been used to determine the reliability of redundant systems that consist of hundreds of independent components.

E.2 *FCASE* `start System` Section

The `start System` section of the *FCASE* input file describes the logical relationships between the elements that compose the system. The variables that define the system may be either vectors that represent sets of redundant components or scalars that represent individual components. *FCASE* includes functions for determining the k-out-of-n:G reliability of redundant sets of components subject to either PFC or IFC. Individual functions are provided for PFC, ELC, FLC and OLC coverage models. *FCASE* uses the symbol & to represent the Boolean "AND" function operator, and it uses the symbol | to represent the Boolean "OR" function operator. The

& and | operators can operate on either vectors or scalars. For vector operations, the operands must, of course, be of equal length. The vector operations are performed pairwise, element by element, for each of the vectors. Thus,

```
a = b & c;
```

defines the variable a as having a reliability equal to the reliability of b **and** c. Also, in a similar fashion,

```
a = b | c;
```

defines the variable a as having a reliability equal to the reliability of b **or** c. Expressions involving the & and | operators can be of arbitrary complexity and can make use of parentheses to define the calculation sequence. The same syntax is used in these cases as would be used in a C or FORTRAN statement; for example,

```
a = (b | c) & (d | e);
```

The *FCASE* functions for determining the *at least* k-out-of-n:G reliability of a set of redundant elements defined by the vector p are

- `PFC(k, p)` for an at least k-out-of-n:G PFC system
- `ELC(k, p, c)` for an at least k-out-of-n:G ELC system
- `FLC(k, p, c)` for an at least k-out-of-n:G FLC system
- `OLC(k, p, c)` for an at least k-out-of-n:G OLC system

The following are *FCASE* functions for determining the *exactly* k-out-of-n:G reliability of a set of redundant elements defined by the vector p:

- `PFCX(k, p)` for an exactly k-out-of-n:G PFC system
- `ELCX(k, p, c)` for an exactly k-out-of-n:G ELC system
- `FLCX(k, p, c)` for an exactly k-out-of-n:G FLC system
- `OLCX(k, p, c)` for an exactly k-out-of-n:G OLC system

The second argument in the list above, p, can also be defined in a compound manner, as with

```
sys = PFC(1, a & b);
```

where a and b are vectors and the 1-out-of-n:G selection is made on their product. Redundant systems may consist of a single k-out-of-n:G set of elements, or they may have several embedded k-out-of-n:G subsystems.

E.3 *FCASE* `start Results` Section

This section summarizes the full set of commands that can be invoked in the `start Results` section of the *FCASE* input file. The full set of *FCASE* commands for this section is

- `Qrepeat(var, tStart, tEnd, deltaT);`
  - Returns a list of times and the probability (1 – `var`) as a function of time (that is, the probability of `var` failure)
  - `var`: any scalar variable defined in the `Start System` section
  - `tStart`: start time for the probability list
  - `tEnd`: end time for the probability list
  - `deltaT`: time increment for the probability list
- `repeat(var, tStart, tEnd, deltaT);`
  - Same as `Qrepeat` above
- `Rrepeat(var, tStart, tEnd, deltaT);`
  - Returns a list of times and the probability `var` as a function of time (that is, the reliability of `var`)
  - `var`: any scalar variable defined in the `Start System` section
  - `tStart`: start time for the probability list
  - `tEnd`: end time for the probability list
  - `deltaT`: time increment for the probability list
- `repeatlist(var, tStart, tEnd, deltaT);`
  - Returns a list of probabilities (1 – `var`) as a function of time (that is, the probability of `var` failure)
  - `var`: any scalar variable defined in the `Start System` section
  - `tStart`: start time for the probability list
  - `tEnd`: end time for the probability list
  - `deltaT`: time increment for the probability list
  - Useful for pasting into *Mathematica* when generating a **`LogLogListPlot`**
- `contribution(var, t);`
  - The contribution of the failure rate of each system component to the probability of `var` at time `t`
  - The contribution ratio is the factor by which overall system reliability would improve if the `var` element had a zero failure rate
- `erl(var, t);`
  - Equivalent redundancy level (ERL) of the system; defined by `var` at time `t`
  - ERL is the slope of the probability of failure curve for `var` as measured on a log-log plot (measured in decades per decade)

- `path(var);`
  - Symbolic representation of the probability of `var`
  - Maximum number of polynomial terms in a symbolic expression is two million
  - Since the symbolic expression for each polynomial term can be a few hundred bytes, this option should be reserved for small systems that are known to have polynomials of reasonable length
- `stat;`
  - Outputs the BDD statistics for the problem defined in the `Start System` section
- `basicvartable;`
  - Table of all input variables used to define the problem
- `vartable;`
  - Table of all input variables and any derived variables
- `probtable;`
  - Table of all commands in the `Start System` section that are expanded from vector to scalar format
  - This command can be useful for diagnostic purposes since this is "code" executed by *FCASE*

In the applicable cases above, `var` can be any scalar variable that is defined in the `Start System` section of the input file. Under normal circumstances, `var` represents the probability of failure for either the entire system that is being defined or for some portion of that system. The user should keep in mind that *FCASE* normally generates the probability that a variable (which represents a system, subsystem or component) is in an operational (good) state; that is, *FCASE* generates a reliability value. The `Qrepeat`, `repeat` and `repeatlist` commands return the probability of `var` failure, and `Rrepeat` returns the probability of `var` (that is, it returns the reliability of `var`).

E.4 Comments on *FCASE* Numerical Precision

The *FCASE* probability calculations are performed using double-precision arithmetic: normally about 16 to 18 significant digits. Although the internal reliability representation that is encoded in a BDD is exact, the probability results are presented numerically. The numerical evaluation of a reliability polynomial that represents a highly reliable system is problematic for a probability of system failure that is less than approximately 10^{-12}, since the results contain only a limited number of significant digits. This difficulty arises because the system reliability p for a highly reliable system is only slightly less than unity, and the reported system unreliability q is then the difference between two floating point numbers that are nearly equal ($q = 1 - p$). For systems with a probability of failure that is less than approximately 10^{-14}, the numerical value can be returned incorrectly as zero. This circumstance can be easily identified when the probability of failure for low mission times is reported as zero, but for higher mission times, the probability is greater than zero.

Index

MIX
Papier aus verantwortungsvollen Quellen
Paper from responsible sources
FSC® C105338

If you have any concerns about our products,
you can contact us on
ProductSafety@springernature.com

In case Publisher is established outside the EU,
the EU authorized representative is:
Springer Nature Customer Service Center GmbH
Europaplatz 3, 69115 Heidelberg, Germany

Printed by Libri Plureos GmbH
in Hamburg, Germany